BEEHIVES AND BEE KEEPERS'APPLIANCES

A PRACTICAL MANUAL FOR
HANDMADE BEE HIVES,
WAX AND HONEY EXTRACTION TOOLS,
AND TRADITIONAL APIARY WORK

BY **PAUL N. HASLUCK**

ORIGINALLY PUBLISHED IN 1911

LEGACY EDITION

HASLUCK'S TRADITIONAL SKILLS LIBRARY
BOOK 1

Doublebit Press
Eugene, OR

New content, introduction, and annotations
Copyright © 2019 by Doublebit Press. All rights reserved.

Doublebit Press is an imprint of Eagle Nest Press
www.doublebitpress.com | Eugene, OR, USA

Original content under the public domain.
Originally published in 1911 by Paul N. Hasluck

This title, along with other Doublebit Press books including the Hasluck's Traditional Skills Library, are available at a volume discount for youth groups, outdoors clubs, or reading groups.

Doublebit Press Legacy Edition ISBNs
Hardcover: 978-1-64389-051-7
Paperback: 978-1-64389-052-4

Disclaimer: Because of its age and historic context, this text could contain content on present-day inappropriate methods, activities, outdated medical information, unsafe chemical and mechanical processes, or culturally and racially insensitive content. Doublebit Press, or its employees, authors, and other affiliates, assume no liability for any actions performed by readers or any damages that might be related to information contained in this book. This text has been published for historical study and for personal literary enrichment toward the goal of preserving the American handcraft tradition, timeless trade skills, and traditional artisanal knowledge.

First Doublebit Press Legacy Edition Printing, 2019

Printed in the United States of America
when purchased at retail in the USA

INTRODUCTION
To The Doublebit Press Legacy Edition

The old experts of artisanal trades, country and homestead knowledge, and the woods and mountains taught timeless principles and skills for centuries. Through their timeless books, the old experts offered rich descriptions of how the world works and encouraged learning through personal experiences *by doing*. Over the last 125 years, manufacturing, farming, and construction have substantially changed. Of course, many things have gotten simpler as equipment and technology have improved. In addition, some activities of pre-digital times are now no longer in vogue, or are even outright considered inappropriate or illegal. However, despite many of the positive changes in manufacturing and crafting methods that have occurred over the years, *there are many other skills and much knowledge that have been forgotten.*

By publishing *The Hasluck Traditional Skills Library*, it is our goal at Doublebit Press to do what we can to preserve and share the works from forgotten teachers that form the cornerstone of the history of the American artisans and traditional crafts. Through remastered reprint editions of timeless classics, perhaps we can regain some of this lost knowledge for future generations.

This book is an important contribution traditional handcraft and country skills literature and has important historical and collector value toward preserving the American handcraft and outdoors tradition. The knowledge it holds is an invaluable reference for practicing skills and hand craft methods. Its chapters thoroughly discuss some of the essential building blocks of knowledge that are fundamental but may have been forgotten as equipment gets fancier and technology gets smarter. In short, this book was chosen for Legacy Edition printing because much of the basic skills and knowledge it contains has been forgotten or put to the wayside in trade for more modern conveniences and methods.

With technology playing a major role in everyday life, sometimes we need to take a step back in time to find those basic building blocks used for gaining mastery – the things that we have luckily not completely lost and has been recorded in books over the last two centuries. These skills aren't forgotten, they've just been shelved. *It's time to unshelve them once again and reclaim the lost knowledge of self-sufficiency.*

Based on this commitment to preserving our outdoors and handcraft artisanal heritage, we have taken great pride in publishing this book as a complete original work. We hope it is worthy of both study and collection by outdoors folk in the modern era of outdoors and traditional skills life.

Unlike many other photocopy reproductions of classic books that are common on the market, this Legacy Edition does not simply place poor photography of old texts on our pages and use error-prone optical scanning or computer-generated text. We want our work to speak for itself, and reflect the quality demanded by our customers who spend their hard-earned money. With this in mind, each Legacy Edition book that has been chosen for publication is carefully remastered from original print books, *with the Doublebit Legacy Edition printed and laid out in the exact way that it was presented at its original publication.* We provide a beautiful, memorable experience that is as true to the original text as best as possible, but with the aid of modern technology to make as beautiful a reading experience as possible for books that can be over a century old.

Because of its age and because it is presented in its original form, the book may contain misspellings, inking errors from print plates, and other printing blemishes that were common for the age. However, these are exactly the things that we feel give the book its character, which we preserved in this Legacy Edition. During digitization, we ensured that each illustration in the text was clean and sharp with the least amount of loss from being copied and digitized as possible. Full-page plate illustrations are presented as they were found, often including the extra blank page that was often behind a plate. For the covers, we use the original cover design to give the book its original feel. We are sure you'll appreciate the fine touches and attention to detail that your Legacy Edition has to offer.

For traditional handcrafters and classic artisanal enthusiasts who demand the best from their equipment, this Doublebit Press Legacy Edition reprint was made with you in mind. Both important and minor details have equally both been accounted for by our publishing staff, down to the cover, font, layout, and images. It is the goal of Doublebit Legacy Edition series to be worthy of collection in any outdoorsperson's library and that can be passed to future generations.

Every book selected to be in this series offers unique views and instruction on important skills, advice, tips, tidbits, anecdotes, stories, and experiences that will enrich the repertoire of any person who enjoys escaping a bit from today's modern technology-based, cookie-cutter, and highly industrialized skills. Instead, folks seeking to make things with their hands like the old days may find great value from these resurrected instructional manuals from the past. These books were not simply written to be shelved in a library – they contain our history and forgotten methods to make things with real character and energy with a *human* component.

Therefore, to learn the most basic building blocks of a craft leads to mastery of all its aspects. We hope this book helps you along this path with its rich descriptions and illustrations!

About Hasluck's Traditional Skills Library

Paul N. Hasluck was a prominent author on artisan skills and traditional handcrafts toward the end of the 19th Century. He was the editor of the magazine *Work*, which was a popular handcraft, shop skills, and artisanal craft magazine of the day. His broad expertise in making things with your hands led him to write or edit over 30 volumes on specific handcrafts, arts, and mechanics, with each manual containing invaluable information related to each craft.

Hasluck had a great eye for collecting the info that beginners and experts alike needed to perfect their craft. His volumes were loaded with helpful diagrams, tables, and illustrations that are useful even by today's digital standards. In short, Hasluck's instructional manuals were the *go-to instructional library* if someone wanted to learn a particular skill. Used by the U.S. military, the Boy and Girl Scouts, and countless folks at farms, public libraries, and homes across the world, Hasluck's instructional manuals were the perfect "handy book" for learning.

This Doublebit Press Legacy Edition republishes this tradition of handcrafted quality and artisanal work. We hope that this deluxe printed edition of this work will help you gain mastery in your craft, as it is presented in the exact form that it was originally published. Even today, the knowledge contained within its pages are timeless and have much to teach!

Finally, as art, Hasluck's manuals contain beautiful illustrations and line art that are a sign of simpler, yet authentic times when quality mattered and craftsmanship was king. This collectible volume makes a great addition to the bookshelf of any handcrafter, maker, artisan, farmer, homesteader, or outdoors enthusiast!

BEEHIVES AND BEE KEEPERS' APPLIANCES

WITH NUMEROUS ENGRAVINGS AND DIAGRAMS

EDITED BY

PAUL N. HASLUCK

Author of "Handybooks for Handicrafts," etc. etc.

PHILADELPHIA

DAVID McKAY, Publisher

610, *SOUTH WASHINGTON SQUARE*

1911

PREFACE.

THIS Handbook contains, in form convenient for everyday use, a comprehensive digest of the knowledge of beehives and bee keepers' appliances, scattered over more than twenty thousand columns of WORK—one of the weekly journals it is my fortune to edit—and supplies concise information on the details of the subjects of which it treats.

Readers who may desire additional information respecting special details of the matters dealt with in this Handbook, or instructions on kindred subjects, should address a question to the Editor of WORK, La Belle Sauvage, London, E.C., so that it may be answered in the columns of that journal.

P. N. HASLUCK.

La Belle Sauvage, London.

CONTENTS.

CHAPTER	PAGE
I.—Introduction: A Bar-frame Beehive	9
II.—Temporary Beehive	19
III.—Tiering Bar-frame Beehive	25
IV.—The "W.B.C." Beehive	41
V.—Furnishing and Stocking a Beehive	48
VI.—Observatory Beehive for Permanent Use	68
VII.—Observatory Beehive for Temporary Use	74
VIII.—Inspection Case for Beehives	84
IX.—Hive for Rearing Queen Bees	89
X.—Super-clearers	96
XI.—Bee Smokers	102
XII.—Honey Extractors	116
XIII.—Wax Extractors	133
XIV.—Bee Keepers' Miscellaneous Appliances	141
Index	158

LIST OF ILLUSTRATIONS.

FIG.	PAGE
1.—Cross Section of Bar-frame Beehive	12
2.—Longitudinal Section of Bar-frame Beehive	13
3.—Floor Board of Beehive	14
4.—Brood Chamber Outer Casing	15
5.—Slide Piece for Entrance	15
6.—Lift over Brood Chamber	16
7.—Cone Bee Escape in Beehive Roof	17
8.—Lining of Brood Chamber	18
9.—Temporary Beehive	19
10.—Section of Temporary Beehive	20
11.—Section of Inner Wall	21
12. Box Cut to Fit Two Roofs	22
13.—Device to Prevent End-shake of Frames	23
14.—Tiering Bar-frame Beehive arranged for Summer Use	27
15.—Tiering Bar-frame Beehive arranged for Winter Use	29
16.—Longitudinal Section of Tiering Bar-frame Beehive	30
17.—Cross Section of Tiering Bar-frame Beehive	31
18.—"Eke" for Increasing Height of Lift	32
19.—Body-box of Tiering Bar-frame Beehive	33
20.—Tongue-and-groove Joint	34
21.—Floor Bearer	35
22.—Floor Board	36
23.—Cutting Beehive Roof Gables	37
24.—Beehive Roof Wings	39
25.—Parts of "W.B.C." Beehive	42
26.—Inside of "W.B.C." Beehive	43
27.—Plinth	44
28.—Section of Top Edge of Front and Back of Body-box	45
29, 30.—Marking Out Legs of Stand	45, 46
31.—Cutting Shoulders of Legs of Stand	46
32, 33.—Cross Sections of Stand Legs	47
34.—B.B.K.A. Standard Frame	43
35.—Broad-shouldered Frame	49
36.—Abbot's Broad-shouldered Frame	49
37, 38.—Plain Frames with "W.B.C." Ends	50
39.—Wired Bar Frame	51
40.—American Jointed Frame	52
41.—Securing Side Bars of Frame to Top Bar	52
42.—Wax Sheet Fixed in Frame	53
43.—Frame Wired to receive Foundation	53
44.—Block for Wiring Frame	54
45.—Gauge Board	55
46.—Woiblet Spur Embedder	56
47.—Wheel of Woiblet Spur Embedder	56
48.—Spur Embedder with Wooden Handle	57
49.—Embedder made with Floor Brad	57
50.—Hoffman Self-spacing Frame	58
51.—"W.B.C." Tinplate End	59
52.—Cast Metal End	59
53.—Howard Tinplate End	60
54.—Pine's Cast Metal End	60
55.—Staples used as End Spacers	61
56.—Back View of Dummy	62
57.—Section of Dummy	63
58.—Simple Form of Dummy	63
59.—Old Style of Queen Excluder	63
60.—Section with Foundation	64
61.—Section before Folding	64
62.—Cross Section of Crate	65
63.—Longitudinal Section of Crate	65
64.—Section of Metal Bar	65
65.—Section Divider	66
66.—Vertical Section through Observatory Beehive	69
67.—Half Cross Section and Half Back Elevation of Observatory Beehive	70
68.—Front Elevation of Observatory Beehive	71
69.—Observatory Beehive for Temporary Use	75

8 BEEHIVES AND BEE KEEPERS' APPLIANCES.

FIG.	PAGE
70.—Vertical Cross Section of Observatory Beehive	76
71.—Horizontal Section of Observatory Beehive	77
72.—Sectional Plan of Observatory Beehive Top	77
73.—Inside Frame of Observatory Beehive	79
74, 75.—Mounting Observatory Beehive	81, 82
76.—Inspection Case for Beehives	84
77.—Section of Inspection Case	85
78.—End Elevation of Inspection Case	85
79.—Arrangement of Case to Fit Two Lengths of Beehives	86
80.—Securing End Openings of Inspection Case	87
81.—Cross Section of Inspection Case	87
82.—Handle for Lifting Frames	88
83.—Hive for Rearing Queen Bees	90
84.—Plan of Body-box	91
85.—Section of Hive for Rearing Queen Bees	91
86.—Modified Body-box	92
87.—Division Board	92
88.—Distance Rack	93
89.—Foot of Hive for Rearing Queen Bees	94
90.—Porter Bee Escape	97
91.—Super-clearer Complete	98
92.—Section of Super-clearer	99
93.—Clearer in Use between Hive and Super	100
94, 95.—Bingham Bee Smoker	103, 104
96.—Pattern of Funnel	105
97.—Smoker Diaphragm	106
98.—Coned Blast Pipe	107
99.—Nicked Tube for Making Blast Pipe	107
100.—Spring	109
101.—Wire for Making Spring	109
102.—Bottom Board of Bellows	110
103.—Clarke Smoker	111
104.—Section of Clarke Smoker	111
105.—Pattern of Funnel	113
106.—Bottom of Funnel	114
107.—Diaphragm of Clarke Smoker	114
108.—Little Wonder Honey Extractor	117
109.—Section of Little Wonder Extractor	118
110.—Section of Can and Cage	118
111, 112.—Little Wonder Can and Cage	119
113.—Pattern for Top and Bottom of Extractor	120
114.—Corner of Cage of Little Wonder Extractor	121
115, 116.—Cylinder Honey Extractor	122
117.—Frames, Baskets, etc.	123
118.—Wired Tinplate	123
119.—Rectangular Band and Bridge	124
120.—Pattern of Slide	125
121.—Bands, Slides, and Bridges of Extractor	127
122.—Half of Comb Basket	128
123.—Section through Comb Basket	129
124.—Bolt for Cross-bar	132
125.—Solar Wax Extractor	134
126.—Glazing Top of Solar Wax Extractor	135
127.—Foot of Solar Extractor Stand	135
128.—Revolving Top of Wax Extractor	136
129.—Washer and Screw	136
130.—Section of Gerster Wax Extractor	137
131.—Pattern for Cylindrical Top of Boiler	139
132–134.—Float Bee Feeder	141
135.—Raynor Bee Feeder	142
136, 137.—Base for Bee Feeder	142, 143
138.—Bee Feeder with Square Base	143
139.—Hone Dummy Feeder	144
140.—Rapid Bee Feeder	145
141.—Section of Rapid Bee Feeder	147
142.—Bennett Self-hiver	148
143.—Alley's Self-hiver	149
144.—Front View of Swarm Catcher	150
145.—Section of Swarm Catcher	151
146.—Hive Entrance with Flexible Springs	151
147, 148.—Queen Cages	152, 153
149.—Spring for Queen Bee Cage	153
150, 151.—Driving Irons	154
152.—Bingham Knife	155
153.—Comb Cutter	155
154.—Cheshire Transferring Board	156
155.—Cutting Tongues from Board	157

BEEHIVES AND
BEE KEEPERS' APPLIANCES.

CHAPTER I.

INTRODUCTION: A BAR-FRAME BEEHIVE.

BEE keeping has long been a science, and it took its greatest step towards that position when the bar-framed hive superseded the primitive straw skep. The bee keeper can now regulate the affairs of the bees' household, arrange their marriages, the strength of their forces, the proportion of males to females, and their comfort in summer and in winter. He partakes of the fruits of their labour, and causes them to obtain far more honey than their natural instinct would prompt. All these powers, and several others, date from the invention and perfection of the bar-framed hive. Before that time, from ten to twenty pounds of honey were consumed by the bees in manufacturing one pound of wax, and as honey and wax were about the same price then, it was evidently a loss of about two thousand per cent. to the bee keeper. The idea struck somebody that if he gave the bees a lump of wax, they might be induced to utilise that and save the honey, but they would not touch it. Then he thought of reducing the wax to thin sheets, but the bees would not touch it even in that condition. The next step was to impress the foundation of the cells upon the sheet of wax, and the bees were found to take kindly to it in this condition, to draw it out into perfect combs for brood and honey. The modern bee keeper, therefore, supplies his bees with sheets of comb foundation, which can be purchased at a cheap rate, and will save at least ten times its cost.

One of the advantages of the movable-frame hive is the facility with which comb foundation can be fixed in it, but it is only a minor advantage, for with a little management foundation can be fixed in almost any hive. The principal advantage is the ease with which all the combs and the health and condition of the inmates can be examined. For example, it is sometimes found that the queen is lost, and if there are no eggs or very young larvæ with which to make a new queen, the fate of the colony is doomed except a new queen is provided, and the bee keeper could scarcely ascertain the loss of the queen were he to use the ordinary skep hives. Again, towards winter some colonies have more than enough stores, while others, if left to themselves, would die of starvation before the spring came; the bee keeper, ascertaining this, gives to the weak ones some of the superfluous stores of the others. This shows not alone the advantage of the movable-frame hive, but the necessity of having all the frames in an apiary interchangeable, and made to a standard size.

The question is often asked by a novice in bee culture, "What is the size of a standard hive?" Whilst there is no absolutely fixed standard in this country, the British Bee Keepers' Association recommend a standard frame, which has been adopted generally throughout the British Isles; its dimensions are 14 in. long by $8\frac{1}{2}$ in. deep, the top bar being 17 in. long, thus forming a lug at each end for convenience of handling the frames when filled with comb.

The British Bee Keepers' Association have also determined the thickness of material of which the frames are to be made—the top bar $\frac{3}{8}$ in. thick, the sides $\frac{1}{4}$ in. thick, and the bottom bar $\frac{1}{8}$ in. thick, all being $\frac{7}{8}$ in. wide; and as frames of these dimensions are stocked by all bee-appliance dealers, there is some advantage gained by sticking closely to standard measurements. But a thickness of only ⅜ in. for the top bar makes a frame weak and likely, when the comb is filled with honey and brood, to sag considerably, and when surplus receptacles are placed over sagging frames a void is left which the bees are not slow to take advantage of to fill with either brace comb or propolis, both very undesirable substances in the position named. Some apiarists have adopted a top bar ½ in. thick, the extra ⅛ in. of thickness adding considerably to its stability; but the plan has this disadvantage, that the frames are not strictly standard, although standard outside measurements are adhered to.

A "bee-space" is, approximately, $\frac{3}{16}$ in., that is to say, a bee can pass between two boards spaced $\frac{3}{16}$ in. apart. If a space is less than this, the bees will fill it with propolis or bee-glue, a sticky resinous substance exuded by and gathered from the buds of trees; if more than ⅜ in., the space will have comb built in it. It is customary, therefore, to allow a free bee-space of ¼ in. round the ends of the frames, and a space of ⅜ in. below them. It will be evident from this that any hive must be at least 14½ in. long and 8⅞ in. deep, and provision must be made for the overhanging top bar.

As bees build their combs about 1½ in. from centre to centre, it follows that the width of the hive will depend upon how many frames it is decided the hive shall accommodate, and this point is determined by the style of hive adopted.

Hives may be roughly divided into two distinct classes—long (or "combination") and tiering. The former is associated with the name of Charles Nash Abbot, who may be considered the father of British bee keeping, and is a hive warmly advocated by some. In it the frames hang parallel with the entrance, the number provided for being generally fifteen. In the tiering hive, the style most generally used, provision is made for ten frames only, this being the number to which it is usually considered expedient to limit the queen, and the frames are usually arranged so that they hang at a right angle to the entrance.

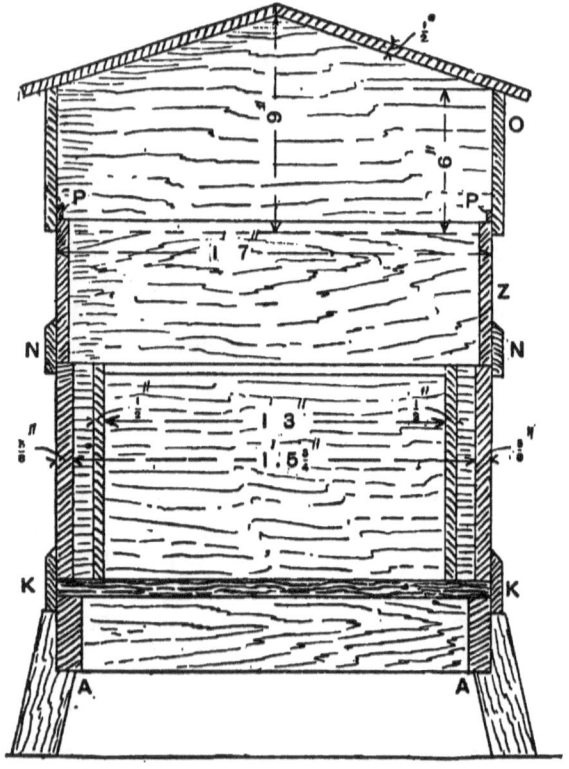

Fig. 1.—Cross Section of Bar-frame Beehive.

The portion of a hive allotted to the use of the queen is spoken of as the "brood nest"; the portions devoted to deprivation purposes are termed surplus chambers or "supers."

The beehive shown in section by Figs. 1 and 2 is of about the best type, as the air space round the brood chamber keeps the bees dry and warm in winter, and cool in summer. It may be worked either for extracted honey in shallow frames, or with two or three crates of sections over the brood nest.

Deal or pine (see p. 25) is used for the whole hive,

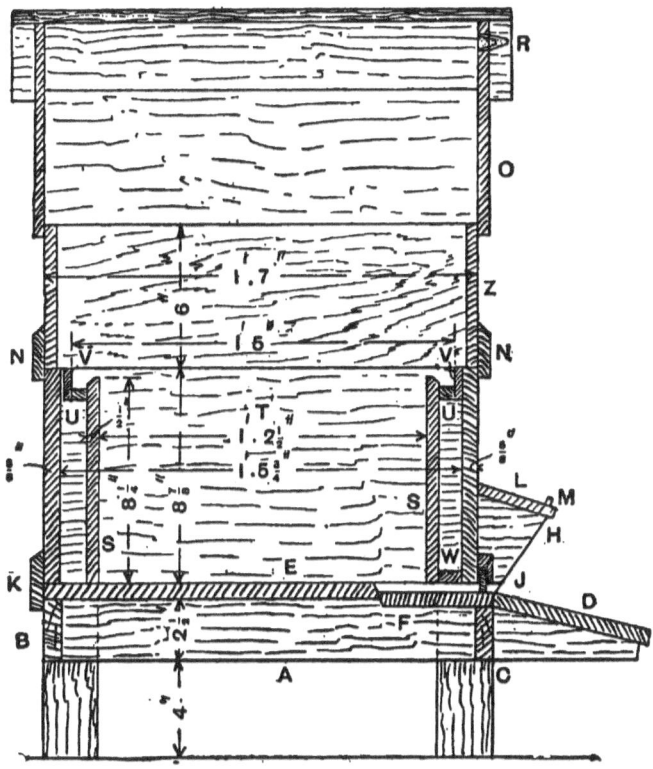

Fig. 2.—Longitudinal Section of Bar-frame Beehive.

and the floor board and stand (Fig. 3) requires two side pieces A (Figs. 1, 2, and 3), each 2 ft. 1 in. long by 2½ in. by 1 in.; one back cross piece B (Fig. 2), 1 ft. 7 in. by 2½ in. by 1 in.; and a front cross piece C, 1 ft. 7 in. by 2⅜ in. by 1 in. Working from the back, the pieces are left the full width for 1 ft. 2 in., then reduced to 2⅜ in. for 5½ in., and at the front sloped down to ¾ in. for the alighting board D (Figs. 2 and 3). The cross pieces are tenoned through the sides, and the back part of the floor E, which should be 1 ft. 7 in. long by 1 ft. 2 in. wide by ⅝ in. thick, is nailed on.

The front part F, 5½ in. wide and ⅜ in. below the level of E to form the entrance, can next be fixed, and then the alighting board D, which is 7 in. wide by ⅝ in. thick, is secured and planed off level with the front of the floor at the joint. Two pieces G (Fig. 3), 5 in. long by 2½ in. by ⅜ in., are prepared and nailed on to form the sides of the entrance, and four legs, 2 in. square and 6 in. long, are cut to the

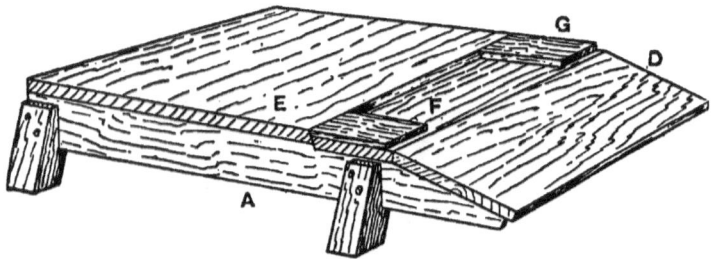

Fig. 3.—Floor Board of Beehive.

shape shown in Figs. 1 and 3, and screwed to the sides.

The outer casing of the brood chamber is shown by Fig. 4, sections of it being given by Figs. 1 and 2. This is made of 9-in. stuff ⅝ in. thick, the front and back pieces being 1 ft. 5¾ in. long, and the sides 1 ft. 10 in. long, 3 in. of which projects in front and is cut to the shape shown at H (Figs. 2 and 4), to carry the porch. Notches J, ½ in. by ¼ in. wide, are for the slides which regulate the size of the entrance.

The box can now be nailed together, but, of course, a better job results from dovetailing or even lapping the joints. A 2½-in. by ½-in. plinth K (Figs. 1, 2, and 4) should project 1 in. below the

INTRODUCTION: BAR-FRAME BEEHIVE. 15

bottom of the box, and is nailed to the sides and back to prevent wet penetrating; it should be bedded in thick white-lead paint, and it is advis-

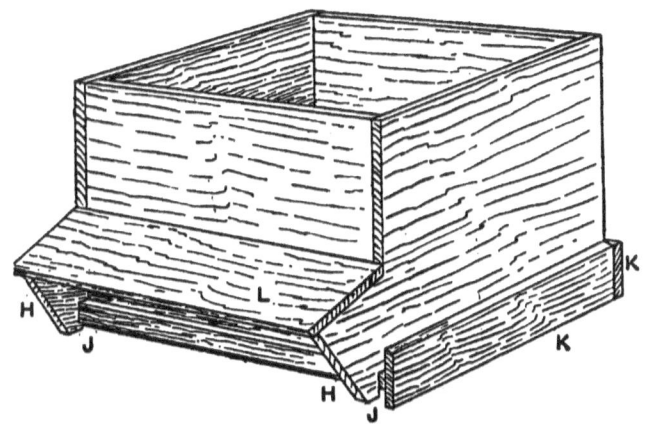

Fig. 4.—Brood Chamber Outer Casing.

able to give all joints a coating of the same material before putting them together.

The porch L (Figs. 2 and 4) is about 3½ in. wide by ½ in. thick, and can be nailed on, the slide piece J at the front being also secured. If the necessary tools for rebating this piece are not available, it may be made up with two pieces, as shown in detail by Fig. 5. Some bee keepers object to a porch sloping towards the alighting board, as

Fig. 5.—Slide Piece for Entrance.

shown in Figs. 2 and 4, as the water drains on to the board, but to prevent this a small strip M (Fig. 2) may be nailed on the full length, or a groove may be cut for the same purpose.

For the lift (Fig. 6) over the brood chamber, two pieces of pine, 1 ft. 6 in. long by 6 in. by ½ in., and two pieces, 1 ft. 7 in. by 6 in. by ½ in., must be prepared. These are nailed together to form a bottomless box, as shown at z (Figs. 1 and 2). The sides of the lift should be made to taper slightly, so that the measurements across the top will be 1 ft. 7 in. bare both ways and 1 ft. $7\frac{1}{16}$ in. across the bottom. This will allow the lift to fit easily over the brood chamber, and the roof to fit easily on to the lift. A plinth N (Figs. 1, 2, and 6), of 2-in.

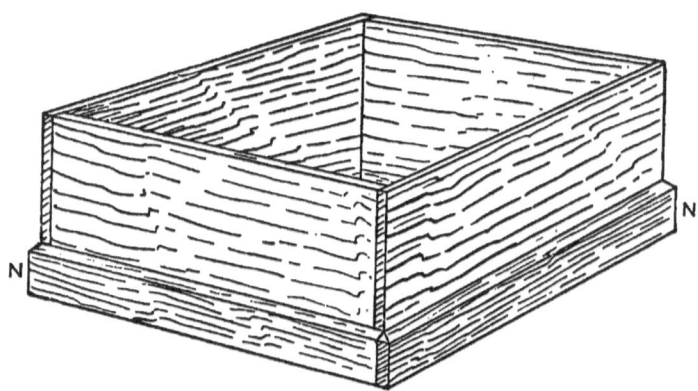

Fig. 6.—Lift over Brood Chamber.

by ½-in. stuff, is nailed round the bottom edge of the lift, to cover the joint.

Each hive should have two of these lifts, to enable an extra lot of sections or shallow frames to be put on, to prevent swarming when honey is very plentiful.

The roof o (Figs. 1 and 2) requires two pieces of ½-in. stuff, 1 ft. 7 in. long by 9 in. wide in the centre, tapered down to 6 in. at both ends, and two pieces, 1 ft. 8 in. long by 6 in. by ½ in., for the sides. These are nailed together and planed level at the top, then covered with two boards each 1 ft. 10 in. long by 1 ft. by ½ in. Over these a covering of sheet zinc should be placed to keep the

INTRODUCTION: BAR-FRAME BEEHIVE.

top weatherproof. To prevent the roof dropping too far over the lift, a couple of ½-in. by ¼-in. strips P (Fig. 1) are nailed inside at the sides, or if a good job has not been made of the fitting, the strips should be fixed all round to keep out robber bees.

For ventilation, a hole is bored near the top of the roof as shown at R (Fig. 2), and this should be fitted with cone bee escapes (Fig. 7), to allow any bees that happen to get over the quilt to find their way out. (Fig. 7 shows a bee escape fitted to another shape of hive.)

The inner lining of the brood chamber is shown by Fig. 8. For this will be required two pieces S

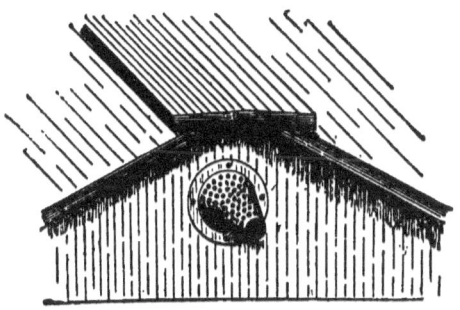

Fig. 7.—Cone Bee Escape in Beehive Roof.

(Figs. 2 and 8) 1 ft. 3 in. long by 8½ in. by ½ in., and two T, 1 ft. 5 in. by 8⅞ in. by ½ in. These are nailed together to leave a space of 1 ft. 2½ in. between the narrow pieces, and then two strips U are prepared, 1 ft. 3 in. by ¾ in. by ½ in., and nailed on, part of Fig. 8 being broken away to show the arrangement of this more clearly. To complete the lining, two strips V, 1 ft. 4 in. by 1¼ in. by ¼ in., are nailed on.

The regulating slides for the entrance consist of two strips 9 in. by 1 in. by ¼ in., and J (Fig. 2) shows one of these in position.

To prevent the bees getting up between the lining of the brood chamber and the casing, a loose

18 *BEEHIVES AND BEE KEEPERS' APPLIANCES.*

piece w (Fig. 2), 1 ft. 4 in. long by 1⅛ in. wide, is dropped in to cover the sinking in the floor board.

The necessary internal fittings for this hive are

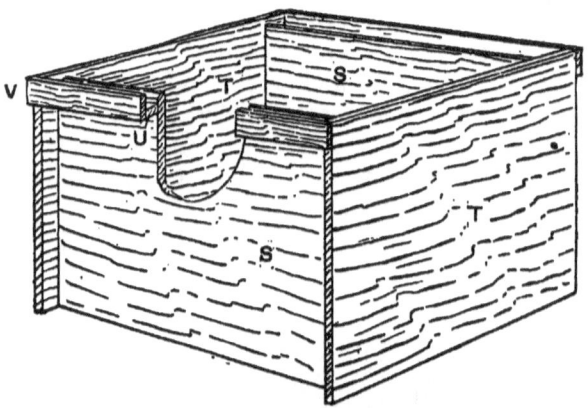

Fig. 8.—Lining of Brood Chamber.

described fully in Chapter V. (pp. 48 to 67). They include bar frames, comb foundation, queen bee excluder zinc, etc.

CHAPTER II.

TEMPORARY BEEHIVE.

WHETHER the apiary be large or small, an appliance which is always in request is a makeshift hive—anything which, whilst not good enough to form a

Fig. 9.—Temporary Beehive.

permanent home for bees, will serve as a temporary lodgment for them until they can be properly housed. Fig. 9 is an isometric view of a hive for such a purpose, and shows an alternative form of porch to that illustrated in the sectional view (Fig. 10).

It is not advisable to waste good material on an article of merely temporary use. Empty packing boxes can generally be obtained from grocers for a few pence each. A couple of Orlando Jones' starch boxes with lids provide sufficient material to complete one hive. These boxes measure, inside, 17 in. long, 13¾ in. wide, and 9½ in. deep, and with but a slight expenditure of labour and material

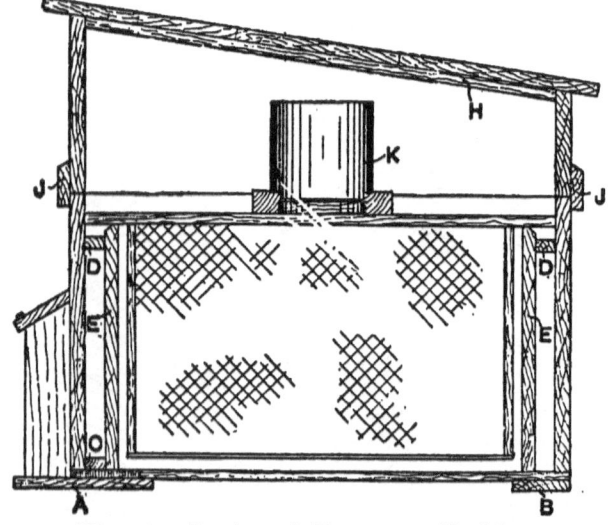

Fig. 10.—Section of Temporary Beehive.

make capital makeshift or "nucleus" hives, holding eight frames, or ten at close spacing.

Select the better of the two boxes, and, choosing the best end for the front, turn it upside down, and mark a line on the bottom parallel with and distant from the front 2½ in., and two others at right angles therewith 3½ in. from each side. Bore a centre-bit hole and, with a pad-saw, remove the parallelogram enclosed by the lines drawn. Now take a piece of the lid, about 4½ in. wide and as long as the width of the box, and nail it across so as to cover the portion of the bottom cut away and to project in front some 2 in. or so, to form an alighting board (see A, Figs. 9 and 10). Nail another narrow strip B across the back end of the box so that it may stand level, and if the joint of the bottom is open, cover it also with a similar strip. Stand the box on its bottom, and it will now be seen, as shown by Fig. 9, that there is an opening formed, about 7 in. wide and ⅜ in. high, by which the bees can enter the hive.

Knock the second box carefully apart, avoiding splitting the boards, and reduce the two ends to 8½ in. in height. Cut them to 8⅝ in. and finish off to

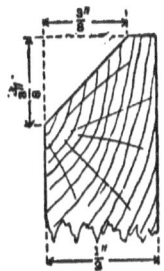

Fig. 11.—Section of Inner Wall.

8½ in. with the plane; then bevel one edge of each as shown by Fig. 11, leaving a flat about ⅛ in. wide on the edge. After having drawn lines on the insides of the hive sides, parallel to and distant from the front and back 1¼ in., insert the reduced end pieces and carefully nail them so that they rest close upon the hive bottom, and their inside faces coincide with the lines drawn.

These inner walls (E, Fig. 10) should then be found 14½ in. distant one from the other, and if a standard frame is inserted it should fit without any end-shake, and there should be a parallel space of ¼ in. between the side bars of the frame and the inner walls of the hive, also a clear space of ⅜ in.

below the frame, as indicated by the section of the hive (Fig. 10).

The bees could now pass in at the entrance and up between the inner and outer walls. To prevent this, three strips of wood, ¾ in. wide and 13¾ in. long (the pieces cut off the inner walls will provide these), are prepared to fit tightly in the vacant space; one of these strips (C, Fig. 10) is pushed down at the front until it rests on the floor of the hive and closes up the open space over the entrance; the remaining two D D are fixed about ½ in. below the top edges of the inner walls E E; this prevents bees getting down whilst the hives are being manipulated, and allows space enough for the fingers to grip the frame ends.

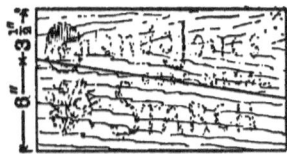

Fig. 12.—Box Cut to Fit Two Roofs.

If a roof is required—it is desirable to provide one—a half-depth starch box may also be obtained, and cut so that the sides slant from front to back, giving a fall of about half the depth of the box—say 2½ in., or thereabouts (Fig. 12 shows how a box may be cut to make two roofs). Nail some of the surplus wood across the top, giving a projection of about 1 in. on all sides, and, after nailing a strengthening piece H inside from front to back, cover the boarding with canvas, calico, felt, linoleum, sheet zinc, or something that, with paint or other substance, can be made impervious to wet. Even a sheet of stout brown paper will suffice if both it and the woodwork are previously well coated with thick paint, and the paper is afterwards given two or more coats of paint.

In order to keep the roof in position on the

hive, plinths J must be prepared, from 1½ in. to 2 in. wide, and nailed to the roof, projecting below its bottom edges about ¾ in. If a porch is wanted, cut out two pieces of wood about 6 in. long and slightly less in width than the projection of the alighting board, slope the top ends, and secure them by nailing through the hive front (do this before the inner walls are put in) and up through the alighting board. A cover board 2½ in. wide is to be fixed to the top (sloped) ends to throw water off, and in order that it may fit close to the hive front the back edge should be bevelled off.

Swarms, also nuclei, sometimes require feeding, so that it is as well to make provision for a feed

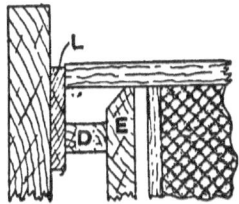

Fig. 13.—Device to Prevent End-shake to the Frames.

bottle K, as shown in position in Fig. 10, by making the roof deeper. In this case another full-depth starch box might be cut across as shown by Fig. 12, when it would furnish two deep roofs. A coat or two of paint will make the hive more sightly and more durable; it will also keep it drier, which is a point of great importance in a permanent hive, though not so serious in one of a makeshift character.

It may be stated that on more than one occasion bees have had to winter in hives similar to the one illustrated in this chapter, and they have made good headway, though they are liable to be severely checked when breeding is in full swing and adverse weather occurs in April and May. These hives, therefore, cannot be recommended for use as permanent quarters for bees. Bees wintered in them also make much greater inroads upon their stores than when housed in proper hives.

Although the particular boxes previously mentioned are recommended, they being of the correct size without alteration, any other box that is at least 17 in. long and not less than 9 in. in depth will serve. If a box is used that allows more than a very little end-shake to the frames—a defect which could not be tolerated—nail a thin strip of wood L on to the inside of each outer wall as shown by Fig. 13; or all the filling-up might, to save trouble, be placed at one end when the extra length is but slight. D E in Fig. 13 agree with D E in Fig. 10.

No entrance slides are here provided; when a contraction of the entrance is required, a strip, or strips, of wood laid on the alighting board will answer every purpose.

CHAPTER III.

TIERING BAR-FRAME BEEHIVE.

TIERING hives are used to afford additional accommodation for brood-raising or honey-storing purposes in cases where the brood nest or hive proper is inadequate. They are made by adding extra bodies above the brood nest.

Beehives are usually made of yellow pine, merely because pine happens to be plentiful in America, and is there used for all common purposes as is deal in this country; consequently pine is used, not because it is the best wood for hives, but because it happens to be cheap. America is the land of bees, honey, and bee keepers, and in that country there are numerous factories exclusively employed, year in, year out, in making beehives and apiarian supplies; and nearly all the beehives made in England are imported from America in the flat—that is, the material is planed and sawn to size on the other side of the Atlantic, and arrives here ready to be put together at a cost very little (if any) above the price at which rough-sawn board of similar material can be purchased in England.

As American winters are so severe, it is the custom above a certain degree of latitude to place the bees in cellars during the cold season—say, from November to March or April—so that hives are seldom exposed to much severe weather, so seldom, in fact, that hive bodies are usually left unpainted. Whilst pine may be employed for the inside parts of English hives, good yellow deal is far preferable for all exposed portions of a beehive, and is considerably cheaper to buy than is first or even second quality yellow pine. Even good quality spruce will be preferable to pine, but it is somewhat liable to twist as it seasons, and is to that extent an undesirable material.

Below is a list of material required for the

LIST OF MATERIAL REQUIRED FOR ONE HIVE.

Number of Pieces.	Net Size in Inches.			Description.	Sides to be Planed.	Letter References in Illustrations.
	Length.	Width.	Thickness.			
	Body Box.					
2	$17\frac{1}{2}$	11	$\frac{5}{8}$	Sides	2	A
2	17	$8\frac{1}{2}$	$\frac{1}{2}$	Inner walls	1	B
1	$18\frac{1}{4}$	$10\frac{7}{8}$	$\frac{1}{2}$	Front	1	C
1	$18\frac{1}{4}$	11	$\frac{1}{2}$	Back	1	D
2	17	$1\frac{1}{2}$	$\frac{1}{4}$	Packing	2	E
3	17	$\frac{3}{4}$	$\frac{1}{2}$	Filling	2	F
	Floor-board.					
2	$20\frac{1}{2}$	2	1	Bearers	1	H
1	17	15	$\frac{1}{2}$	Main floor	1	J
1	17	4	$\frac{5}{8}$	Floor under entrance	1	K
1	17	3	$\frac{5}{8}$	Alighting board	1	L
2	$2\frac{3}{4}$	$2\frac{3}{4}$	$\frac{1}{2}$	Packing blocks	2	M
	Lift.					
2	$18\frac{5}{8}$	11	$\frac{1}{2}$	Sides	2	N
2	$19\frac{3}{8}$	11	$\frac{1}{2}$	Ends	2	P
2	$18\frac{5}{8}$	$\frac{1}{2}$	$\frac{3}{8}$	Fillets	2	R
2	$17\frac{5}{8}$	$\frac{1}{2}$	$\frac{3}{8}$			
	Roof.					
2	$20\frac{1}{2}$	$4\frac{3}{4}$	$\frac{1}{2}$	Gables	2	S
2	$19\frac{1}{4}$	2	$\frac{1}{2}$	Sides	2	T
2	$23\frac{1}{4}$	11	$\frac{1}{2}$	Wings	2	V
1	$23\frac{1}{4}$	$4\frac{1}{2}$	$\frac{1}{2}$	Ridge cover	2	W
2	$19\frac{1}{4}$	$\frac{1}{2}$	$\frac{3}{8}$	Fillets	2	R
2	$18\frac{3}{4}$	$\frac{1}{2}$	$\frac{3}{8}$			
	Porch.					
1	$18\frac{1}{4}$	4	$\frac{3}{8}$	Back	1	P A
2	4	$2\frac{1}{2}$	$\frac{1}{2}$	Brackets	2	P B
1	$18\frac{1}{4}$	1	$\frac{1}{2}$	Distance fillet	1	P C
1	$18\frac{1}{4}$	3	$\frac{1}{2}$	Slope	2	P D
2	7	$1\frac{1}{4}$	$\frac{3}{8}$	Entrance slides	2	P F

making of one tiering bar-frame hive, although it will be found very advantageous to make up at least three at the same time, as each hive is then more likely to be exactly like its fellow, and all parts will interchange—a point of no little importance when working a number of hives. Material

Fig. 14.—Tiering Bar-frame Beehive arranged for Summer Use.

for shallow-frame boxes or section racks is not included herewith, as the making of these and the hive furniture will form the subject of another chapter.

The quantities of sawn boards to order are: 3 ft. of 11 in. by ⅝ in. (¾-in. wrought board will do if sound and not too knotty), 17 ft. of 11 in. by ½ in., 10 ft. of 11 in. by ⅜ in., and 4 ft. of 2 in. by 1 in.

slating batten. These sizes allow for slight waste in trimming, but nothing for waste caused by shakes or large knots. It will be presumed that all the material is cut to size and planed as indicated by the table, and edges and ends squared and shot true.

An isometric view of the hive arranged for summer use is shown by Fig. 14; it consists of floor-board F B, hive-body or brood-nest B N, invertible lift I L, roof R F, and detachable porch D P. Fig. 15 is an isometric view of the same hive arranged for winter with the lift inverted, thus forming what is practically a triple-walled hive. This illustration also shows the hive mounted upon a rough stand which is a condensed milk box, in size 19 in. by 13 in. by 7 in. deep, outside measurements. These boxes are strong, and answer the purpose capitally; and if given a coat of hot tar before being put into use, they will last for years.

The projection in front of the hive is a rough alighting board, the use of which is to keep the space immediately in front of the hive clear, and to prevent tired home-coming bees from being blown upon the ground or grass in front of the hive where, in cold, wet, and windy weather, many are liable to fall to rise no more. These rough alighting boards are from 15 in. to 18 in. (more or less) long, front to back, and in width the width of the floor-board; they can be made of the roughest material—the rougher the better, as affording a firmer foothold for the bees—secured to the underside of fillets about 1 in. by $\frac{3}{4}$ in. in section; these fillets project about an inch above the boarding so as to rest on the floor-board of the hive. A couple of notches should be cut in the underside of the bottom fillet to allow water to drain off, and the board may be left rough, not painted or coated in any way.

Fig. 16 is a longitudinal section of the hive shown by Fig. 14, showing two shallow-frame boxes above the brood-nest; and Fig. 17 is a cross section of the arrangement given in Fig. 15, the dotted lines continued upward indicating the arrangement of the hive for summer use, but showing three section racks above the brood nest instead of the two shallow-frame boxes, as in Fig. 16. As will be seen,

Fig. 15.—Tiering Bar-frame Beehive arranged for Winter Use.

plenty of room is allowed for winter packing, and accommodation is also provided for a stimulating feeder for spring use, as indicated by dotted lines, above which there is still ample room for warm wraps.

The hive is suited alike to those who work solely for honey in the comb, and to those who extract the surplus for bottling; and when, even though it is desired to produce comb-honey as the main crop, a shallow-frame box is added early in the season and left on the hive until its close, room

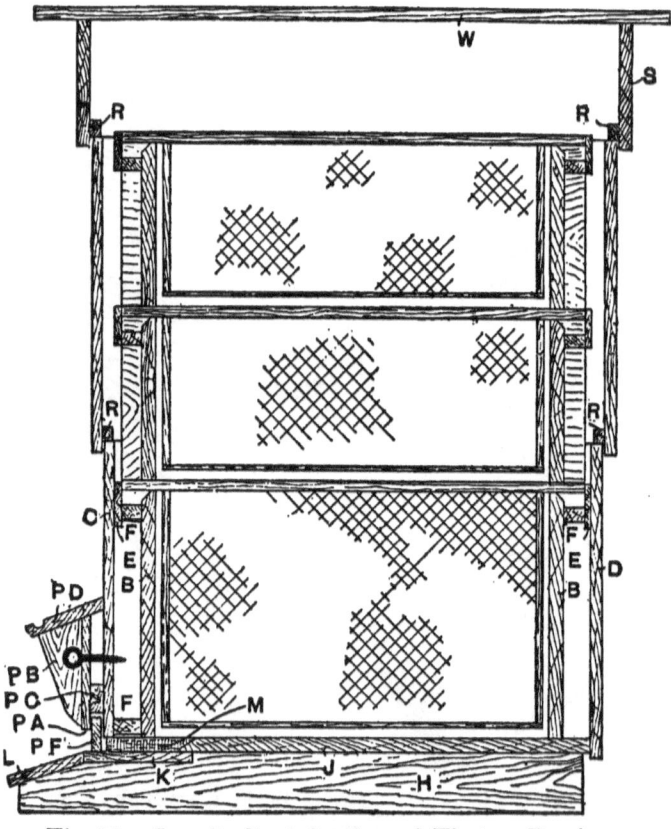

Fig. 16.—Longitudinal Section of Tiering Bar-frame Beehive.

for two section racks above can still be provided by adding an "eke" of the same dimensions as the lift, but about 3 in. deep, with fillets, 1¼ in. wide, projecting ½ in. below its bottom edge on two sides to keep it in position, as in Fig. 18, which shows the underside of the "eke."

TIERING BAR-FRAME BEEHIVE. 31

The body-box is double-walled back and front, and internally it provides accommodation for ten frames at the regulation spacing ($1\frac{9}{20}$ in. centre to centre), two spacing slips P S, and two dummies

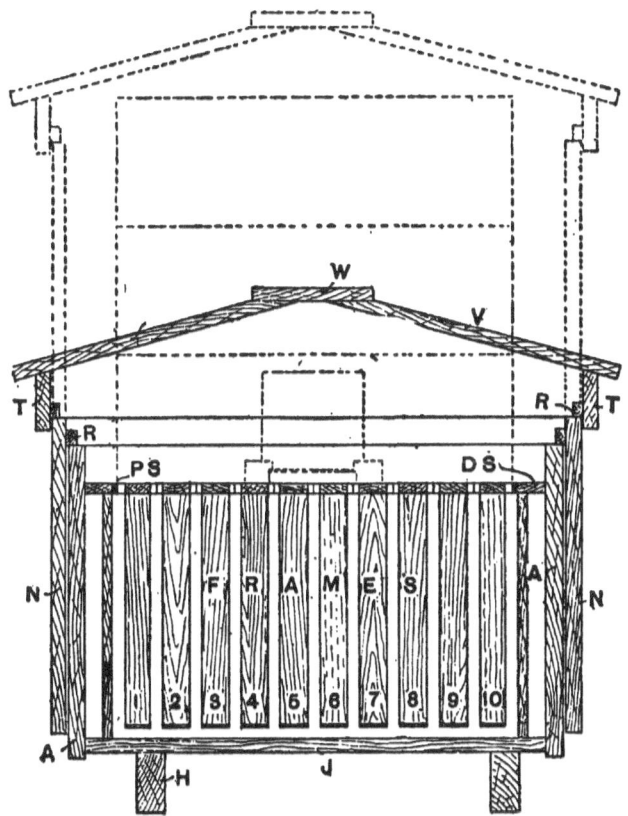

Fig. 17.—Cross Section of Tiering Bar-frame Beehive.

D S (Fig. 17), by means of which the body-box is practically double-walled at the sides also.

Beginning with the body-box, take the two sides A and square across from the top edge of each on the inner side a line distant $1\frac{1}{2}$ in. from each end; the distance between these ($14\frac{1}{2}$ in.) is the internal dimension of the hive, front to back, and allows a ¼-in. space at each end of the frames. Next take the inner walls B, and having bevelled the upper edge of each on the unplaned side (see Fig. 11, p. 21), nail the sides to them, keeping them lower edges ½ in. up from the bottom edges of the sides, and making the planed faces coincide exactly with the lines marked thereon, as shown by Fig. 19.

As the internal dimensions of the hive, front to back, are important, it will be advisable to prepare a gauge-stick—a piece of wood about 1 in. by ½ in. cut truly to length—with which to keep the

Fig. 18.—"Eke" for Increasing Height of Lift.

inner walls the correct distance apart whilst nailing; it is also very important that the inner walls are fixed at right angles to the sides. Measure the distance from the top edges of the sides to the upper edges of the inner walls, and, if all measurements do not coincide, make them do so by planing away the sides where they are highest. As shooting the edges of the sides A will have somewhat reduced their width, the front and back boards, C and D, must also be reduced in width to correspond, and should have the inner face of each smoothed with the plane for about 2 in. down from the top edge. Deduct $\frac{5}{16}$ in. from the measurement of the distance from top of side A to top of inner wall B, already taken, and mark the reduced measurement upon the inner face of both front and back from the top, and below these lines attach with brads the packing strips E. Nail front and back to the ends of the sides, with one nail each

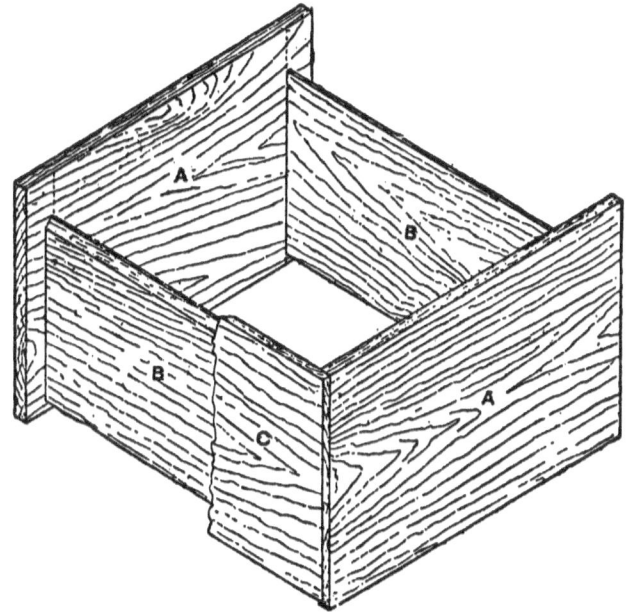

Fig. 19.—Body-box of Tiering Bar-frame Beehive.

at top and bottom, but do not at present drive them home.

Now prepare another gauge-stick fully 17 in. long —say $17\frac{1}{32}$ in.; and if this gauge will not lie comfortably across the inner walls and between the outer walls without tightness or end-shake, distances must be adjusted until it will do so. If it is a tight fit, take a shaving or two as requisite off the face of each packing strip; if the fit is loose, one or two thicknesses of stout paper interposed between front or back and the strips may be sufficient to adjust matters. Measurements being satisfactory, nail on the boards permanently, keeping the front ½ in. up from the bottom edges of the sides, as shown by Fig. 19, clean off the ends flush with the sides, and punch in the nail-heads.

In order to prevent the bees having access to the space between the inner and outer walls, filling strips F should be inserted and secured with brads ½ in. below the top edge of each inner wall, and ⅛ in. above the bottom edges of inner and outer walls at the front. A similar strip may be fixed at the back if desired, but as the bees cannot gain access to the space from the bottom, its provision is superfluous. The body-box is now finished. Oval wire brads for hive work, those for nailing body-

Fig. 20.—Tongue-and-groove Joint.

boxes and lifts to be 2 in. long, are recommended.

The floor-board will next receive attention. It will be noticed in the table of materials (see p. 26) that the main floor is 15 in. wide, consequently a joint will be necessary; and as the floor-board will be more or less exposed to the weather, a glued joint will not stand for any length of time. A tongue-and-groove joint, as shown by Fig. 20, is most suitable, but failing the possession of suitable tools wherewith to work such a joint, a dowelled joint will be the best substitute.

With a gauge set to half the thickness of the wood, mark a line down the centre of the edge of one of the pieces, and make holes with a bradawl on this line 2 in. to 3 in. apart; and, having cut the heads off a sufficient number of wire nails, drive them into the holes with the points projecting about ½ in. Offer the edge of the second board, against the nail points, on a level surface, and make other holes where indicated to receive the points. Coat the edges with thick paint and drive the boards together. Cut the two bearers H to shape as shown by Fig. 21, and, in the sinking cut towards the front, nail the "floor under entrance" K, allowing it to project (say) 2 in. beyond each bearer.

Next nail on the "main floor" J, allowing it to project over the "under" floor ½ in., and the same distance over the ends of the bearers. Bevel one edge of the alighting board L, nail it in place against the "under" floor, and drive a couple of 1¼-in. brads through the face of the alighting board, on the skew, into the edge of the "under" floor.

A reference to Fig. 16 will show that the main floor stops short of the front inner wall of the hive.

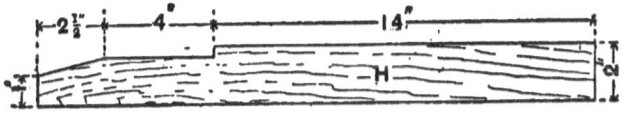

Fig. 21.—Floor Bearer.

This is to provide a means of ingress and egress for the bees; but in order to restrict the entrance somewhat, and also to afford a firm base upon which the body-box may stand, a packing block M is fixed on the "under" floor at each side. Fig. 22 shows the floor-board with one block in position. Bed the blocks in paint, nail them through the "under" floor, and clinch the nails underneath; chamfer the front edge of the main floor between the blocks as shown in Figs. 16 and 22, and the floor-board is finished; but before it is put aside it should be seen that it fits comfortably between the side walls of the body-box, remembering also to make allowance for the thickness of several coats of paint.

The utility of keeping the front of the body-box, also the inner walls, ½ in. above the bottom edges of the sides will now be apparent—the sides and back drop over the floor-board, breaking the joint, keeping the body-box in place, and rendering the provision of plinths unnecessary.

Although making the floor as detailed and illustrated entails a little extra work, the result is worth the pains, as it will be practically an impossibility for wet to drive in on to the main floor; rain might penetrate to the sunk floor, but it could not pass the barrier raised by the higher back portion, consequently the floor would be always dry. Matchboarding might be used for the floor, wrought side

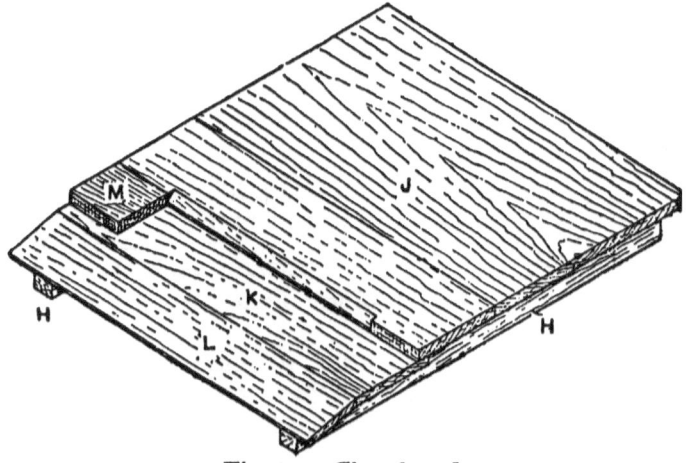

Fig. 22.—Floor-board.

downwards, but the presence of so many joints would give the bees a fine opportunity to daub the floor with propolis.

The lift N scarcely needs description. It is simply a lidless and bottomless box, in size, internally, $\frac{1}{8}$ in. larger each way than the outside measurements of the body-box, over which it is required to telescope. Make sure that it is quite square, and put half a dozen 2-in. nails in each joint so that it will stand a fair amount of rough usage. Fix the fillets R $\frac{1}{2}$ in. from either the top or bottom of the lift to form a stop upon which it may rest in either position upon the body-box, and also to break joint so as to exclude draughts and small vermin.

To make the roof, nail the gables s to the sides T, after having bevelled the upper edges of the latter to correspond with the slope given to the

Fig. 23.—Cutting Beehive Roof Gables.

wings. In cutting out the gables there will be a certain waste of material if one pair only is required; but where not less than three pairs are cut out at one time, they can be cut from an 11-in. board with very little waste, as shown by Fig. 23, the shaded portions representing the "waste."

On each side of the centre line of each gable

draw a line parallel thereto and distant ½ in.; then temporarily tack on the wings v flush with one gable, and with a straightedge draw a line across both gable and wings, as shown by Fig. 24. Then set a bevel to the lines marked, and plane away the top edges of the wings as required. This done, permanently nail the wings in position, allowing them to project 1½ in. over each gable. Cut away the point of each gable level with the bevelled edges of the wings, and nail on the cover-board w, first coating its underside, also each bevel, with thick paint. Well nail each joint at 3-in. intervals, and clinch the nails inside, or screw the joints together from the underside; and if the roof covering is made of seasoned material and is kept painted, no wet will ever find its way inside the hive.

The rim of the roof should be ⅛ in. larger each way inside than the outer dimensions of the lift, and when the fillets R are tacked on inside, ½ in. up from the bottom edge, the roof is finished.

For the purpose of providing a means of egress to stray bees which may linger round the hive top or surplus boxes after manipulations, and also for use at certain seasons as a super clearer, it is well to fit each roof with a cone, as shown by Fig. 7, p. 17. A 1-in. hole should be bored in the front gable previous to putting the roof together, over which, after the hive is painted, a brass perforated cone, which can be purchased for 1½d., is fixed with brass escutcheon pins or small round-headed screws.

The utility of a porch P D is, with some, a matter for argument, it being contended that the advantage it affords in sheltering the entrance to the hive from rain-storms is quite overshadowed by its disadvantage at certain times, as, for instance, when hiving a swarm. As the porch of this hive is detachable, it is not open to any such objection; a few turns of a couple of screw-eyes suffice to remove it from the body-box, and a few more turns will refix it, either in its original position or upon the lift when the latter is telescoped over the body-box at the approach of winter.

Fig. 16 shows the shape of the porch brackets P B which are to be fixed 2 in. in from the ends of the back P A after the top edge of the latter has been bevelled to correspond with the slope of the tops of the brackets. Nail on the distance fillet P C ¾ in. from the bottom edge of the back, and, after bevelling the back edge of the slope so that the top back edge fits close against the hive front when the porch is in position, nail it to both the back and the two brackets, allowing a projection of ½ in. over the back edge of the former. A couple of 1½-in. screw-eyes—brass will be preferable—with a washer under the head of each, will provide an

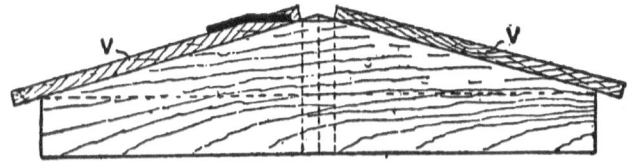

Fig. 24.—Beehive Roof Wings.

easy means of securing the porch, and they should be inserted outside the brackets so as to be easily got at without disturbing the occupants of the hive.

The entrance-slides P F, with which to contract the entrance formed in the floor-board, are simply pieces of wood rectangular in section with the ends cut square, and having a small screw-eye or round-headed screw inserted in each as a handle to assist in drawing it backward and forward in the recess formed at the back of the porch. When entirely withdrawn, the slides may be placed in the open space behind the back of the porch above the distance fillet.

Punch in all nail heads, and give the hive at least three coats of some light-coloured paint, stopping the nail holes with putty between the first and second coats. Although the inside of the hive need not be painted, the floor-board should be painted all over; and it will considerably add to the wear of the hive if all joints are thickly coated with paint before being nailed together.

It is highly important that measurements be strictly adhered to, especially as regards the internal dimensions of the body-box, its squareness being no less essential. Bees actively resent interference when brace combs have to be torn asunder, or when propolised surfaces part with a "snap"; and although accuracy is not so essential in those parts of the hive to which the bees are not allowed access, a properly made and well-fitting hive means comfortable manipulations and few or no stings.

CHAPTER IV.

THE "W.B.C." BEEHIVE.

THE "W.B.C." hive is a bar-frame hive designed by Mr. W. Broughton Carr, and its several parts are shown in the illustrations accompanying this chapter. The stand, floor-board, outer case, lift, and roof are shown in their relative positions by Fig. 25, but drawn apart to illustrate their construction better. Fig. 26 shows the interior parts or hive proper—the eke, body-box or brood-chamber, and a shallow frame-box or super.

The following particulars of the sizes of wood used in making this hive were published by Mr. Carr some years since. The floor-board is 1 ft. 8 in. from front to back, the alighting-board projecting 7 in. as shown. The width of the entrance is 1 ft. $3\frac{5}{8}$ in. by $\frac{1}{2}$ in. high, and the full width of the floor-board is 1 ft. $6\frac{7}{8}$ in., as shown in Fig. 25. The wood of the floor-board is $\frac{1}{2}$ in. thick (the joints being tongued and grooved), and nailed to battens $2\frac{1}{2}$ in. deep by $1\frac{1}{2}$ in. wide, cut as shown.

The front and back boards of the outer case are $\frac{3}{8}$ in. thick by 1 ft. $6\frac{7}{8}$ in. by $8\frac{7}{8}$ in. The side pieces are $\frac{1}{2}$ in. thick by 1 ft. $7\frac{1}{4}$ in. by $8\frac{7}{8}$ in. The inside measurement of the case when nailed up is 1 ft. $5\frac{7}{8}$ in. by 1 ft. $7\frac{1}{4}$ in. A plinth $1\frac{1}{2}$ in. wide is nailed round the lower edge of the case, and drops $\frac{1}{2}$ in., as shown in Fig. 27; this figure also shows a rebate $\frac{1}{8}$ in. by $\frac{1}{2}$ in. taken out of the strips used for the plinth, to fit the case over the floor-board.

The construction of the porch, which is nailed to the front of the case, will be best understood from Fig. 25. The roof of the porch is $4\frac{1}{2}$ in. wide, and on the under-side of the lower edge a groove

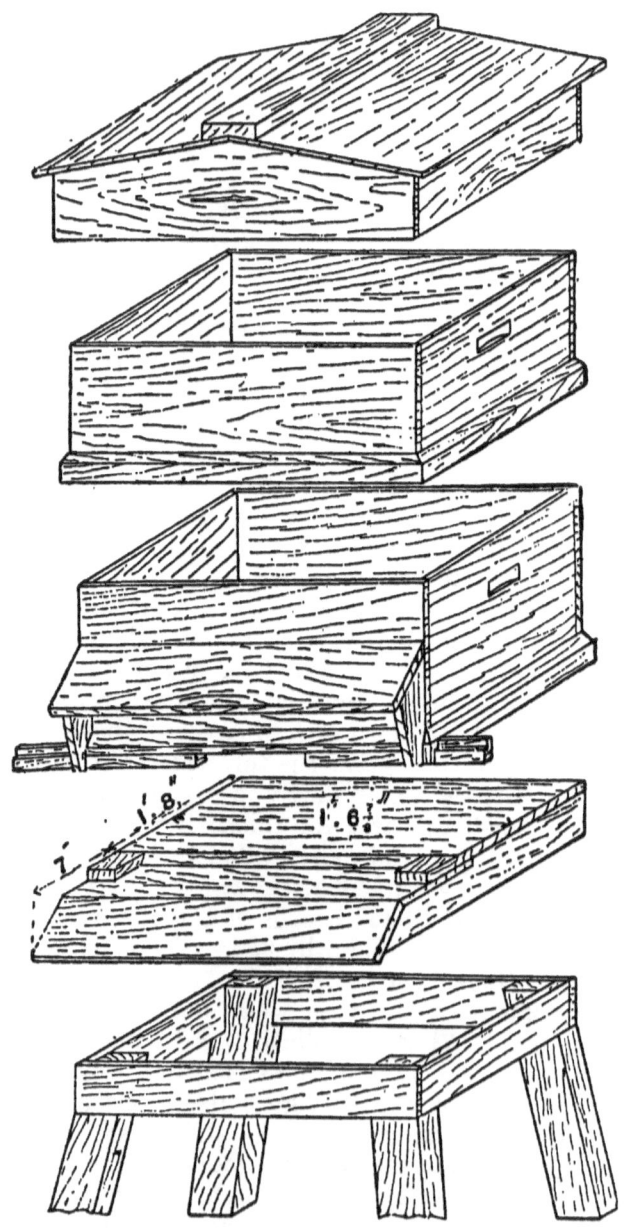

Fig. 25.—Parts of "W.B.C." Beehive.

is cut to turn rainwater off. The entrance can be closed by slides of ½-in. wood, 10 in. long by 1¼ in. wide, rebated along the top edge to slip under the rebated edge of the guide-piece above the entrance.

The lift, which rests on the outer case, is of

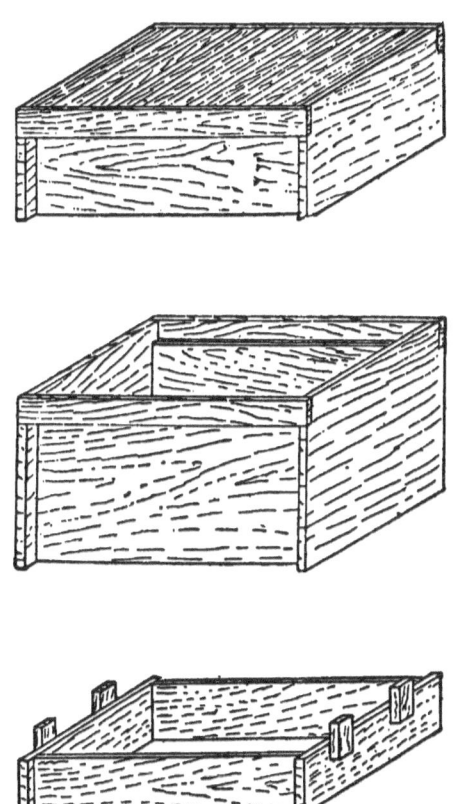

Fig. 26.—Inside of "W.B.C." Beehive.

exactly the same construction as the case, but is only 6½ in. deep.

The roof just slips over the lift or outer case. The front and back pieces are 1 ft. 7¾ in. long, 3½ in. deep at the ends and 5⅜ in. at the centre, to form the ridge, and are of ⅜-in. stuff. The sides are of

¾-in. stuff, 1 ft. 8⅛ in. by 3⅝ in. The lower inner edge of the side pieces is rebated ⅜ in. by ⅜ in., so that the roof may rest on the edge of the case below.

The body-box or brood chamber (Fig. 26) is constructed to hold ten standard frames, and is 1 ft. 2½ in. by 1 ft. 3 in., inside measurement. The front and back pieces are ⅜ in. thick, 15¼ in. long, and 9¼ in. wide, and fit in grooves in the side pieces 1 in. from their ends. The side pieces are 1 ft. 5 in. by 9 in., and ½ in. thick. From the top corners of

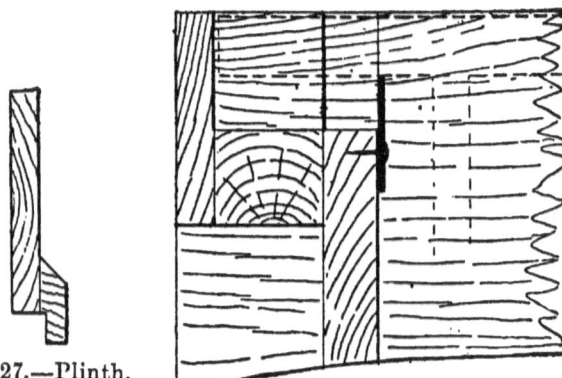

Fig. 27.—Plinth.

Fig. 28.—Section of Top Edge of Front and Back of Body-box.

the side pieces a piece is nicked out to receive strips of wood 1 ft. 4 in. by 1⅜ in. by ¼ in., which extend from side to side at the back and front. These strips enclose the top of the bar-frame ends, keeping them in position. A slip 1 ft. 3 in. long and ¾ in. by ⅝ in. square is nailed between the strips and the front and back pieces of the box. A strip of zinc, on which the frame ends rest, is nailed on the top edge of the front and back pieces. Fig. 28 is a section across the top front and the back edge of the body-box. The dotted lines show a corner of a frame resting on the zinc strip.

The shallow frame-box or super, which fits over

the body-box, is, except that it is 6 in. deep, exactly the same as the body-box. The bar-frames to fit are 5½ in. deep.

The eke, for winter use only, is 3 in. deep, and goes below the body-box. The four cleats nailed to its top edge, shown in Fig. 26, are to keep it in position under the body-box. The eke is not essential, but is used to raise the body-box to give bottom ventilation when wintering the bees. It can also be placed below the shallow frame-box to bring it to the size of the body-box.

The stand for the hive is simple in construction, the only difficult part being in marking out the

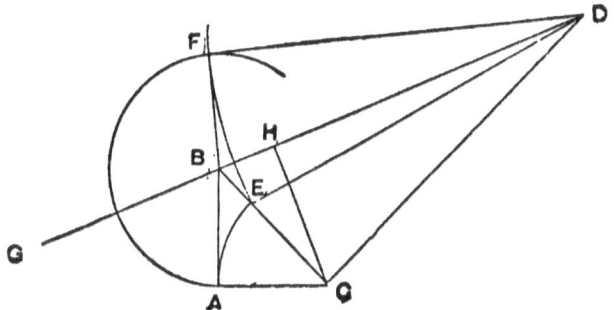

Fig. 29.—Marking Out Legs of Stand.

wood for the legs, which are splayed outwards in the direction of the diagonals of the frame, to which they are fixed. The splay of the legs depends on the depth and thickness of the wood of the frame; and in order that the outside faces of the legs should meet the bottom edge of the frame corners exactly, the cross section of the legs must have the form of a trapezium.

Figs. 29 and 30 show how to find the correct angles to which to set a bevel for marking out the shoulders of the legs. From a point A (Fig. 29), draw two lines A B and A C at right angles to each other, making them equal or proportional to the thickness of the wood of the stand frame—that is

¾ in. Join B C, and from the point C draw the line C D at right angles to B C, making C D equal or proportional to the depth of the frame—that is

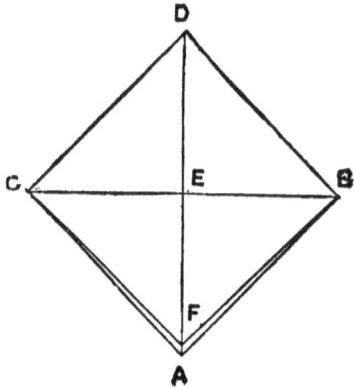

Fig. 30.—Marking Out Legs of Stand.

2½ in. Then join B D, and on the line B C take from the point C a distance C E equal to A C and join E D. With B as centre and A B as radius describe a circle, and with D as centre and D E as radius describe the arc E F cutting the circle at the point F. Join B F, and produce D B in the direction G, then the angle F B G is the angle for the bevel. With the sizes given this angle is about 106°.

The angle for the outside faces of the leg is

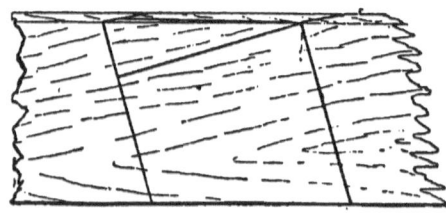

Fig. 31.—Cutting Shoulders of Legs of Stand.

found as in Fig. 30. Draw two straight lines A B and A C at right angles to each other, making them equal to twice the thickness of the wood of the

THE "W.B.C." BEEHIVE. 47

frame—that is 1½ in. Complete the square A B D C, and draw the diagonals. Set off from the point of intersection E a distance E F equal to the perpendicular C H (Fig. 29), and join C F and F B. The angle C F B, about 95°, is the angle to which two adjacent faces of the wood from which the legs

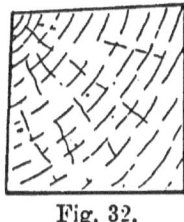
Fig. 32.

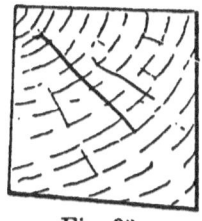
Fig. 33.

Figs. 32 and 33.—Cross Sections of Stand Legs

are to be cut should be dressed before the shoulders are marked on them. Fig. 31 shows the marking out of the shoulder end upon the wood. Fig. 32 is a cross section, and Fig. 33 a section after cutting along the lines A B and B C (Fig. 30).

CHAPTER V.

FURNISHING AND STOCKING A BEEHIVE.

BEING in possession of a hive, the bee keeper must turn his thoughts towards the furnishing of it suitably for the habitation of the bees. True, if given the empty hive, the bees will themselves proceed to furnish it, but most probably in a style quite at variance with the ideas of modern bee keepers.

In a bar-frame or movable-comb hive, it is of great importance that each comb should be built

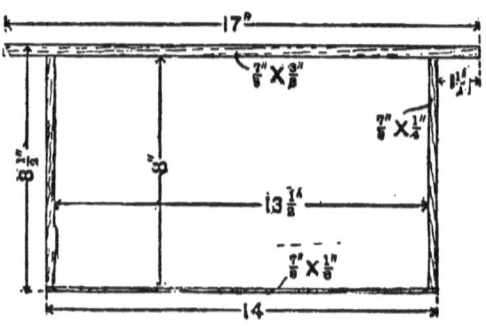

Fig. 34.—B.B.K.A. Standard Frame.

quite straight in its frame, and that each frame should be truly square and in fit condition for being lifted from the hive without tearing asunder any attachment either to another comb or to the hive walls; and this condition can be secured only by correct initial management. It is an old truism that "bees do nothing invariably"; but as a general rule, if they are properly started in the way they should go, they will not make any serious departure from it.

Furnishing and Stocking a Beehive. 49

In furnishing a beehive, the first requisite is a set of frames, each of which will eventually contain a comb. The "standard" frame of the British Bee Keepers' Association is illustrated by Fig. 34, and described on p. 10, and if the measurements there given are adhered to, the pattern of the

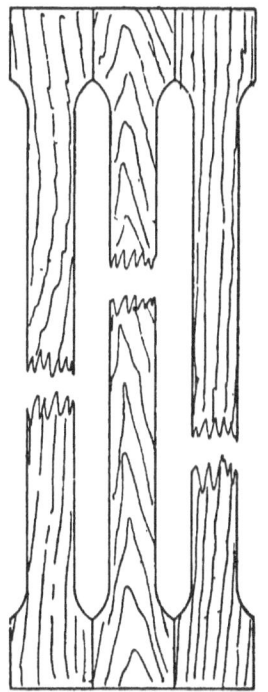
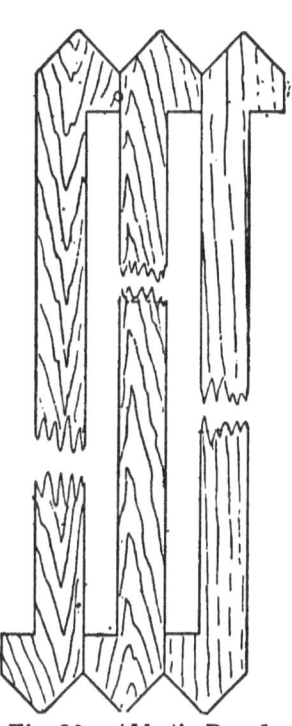

Fig. 35.—Broad-shouldered Frame.

Fig. 36.—Abbot's Broad-shouldered Frame.

frames does not much matter; it will vary according to the method employed of spacing the frames in the hive. Bees in a natural state build their combs from $1\frac{1}{4}$ in. to $1\frac{1}{2}$ in. apart centre to centre, and the spacing usually adopted in placing frames is $1\frac{9}{20}$ in. centre to centre, ten frames occupying a width of $14\frac{1}{2}$ in.

Frames may be roughly divided into two classes:

D

50 *BEEHIVES AND BEE KEEPERS' APPLIANCES.*

broad-shouldered, which are self-spacing; and shoulderless, which require some mechanical contrivance to ensure correct spacing. The latter is the class most commonly employed, although a considerable number of apiarists still hold to the broad-shouldered frame, sometimes called the

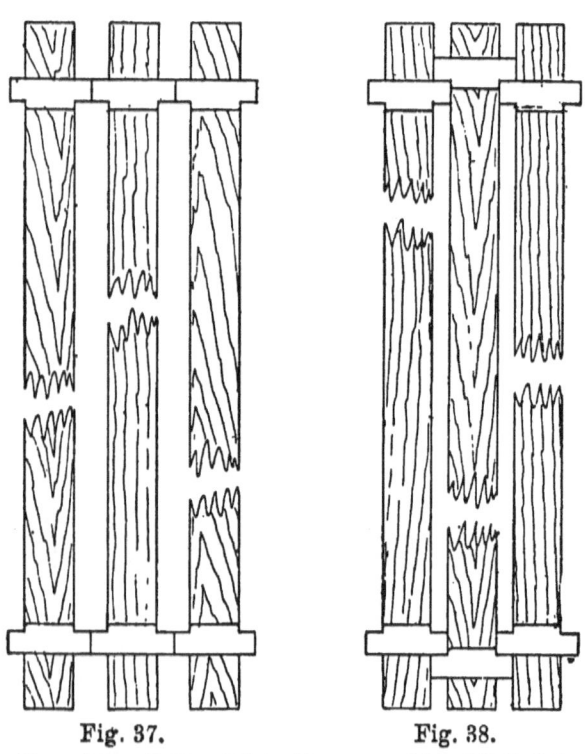

Fig. 37. Fig. 38.

Figs. 37 and 38.—Plain Frames with "W.B.C." Ends, Ordinary and Narrow Spacing.

"Abbot" frame. Figs. 35 and 36 illustrate in plan different forms of broad-shouldered frames, Fig. 36 being the "Abbot" pattern; and Fig. 37 illustrates, also in plan, a shoulderless frame fitted with "W.B.C." metal ends. Fig. 38 illustrates a variation in the working of the "W.B.C." end which will be referred to later.

A plain, shoulderless frame, such as has been previously referred to as the B.B.K.A. "standard" frame, consists merely of the separate pieces of wood sawn or planed to the correct dimensions and squarely nailed together, and for this purpose—as a time saver—a frame block is a convenience upon which the separate pieces composing the frame are laid and held in position whilst they are nailed. Ten such frames (Fig. 39) are required for the hive shown by Figs. 1 and 2, pp. 12 and 13. It is very important that they should be of standard size, with the top bar A 1 ft. 5 in. by $\frac{7}{8}$ in. by $\frac{3}{8}$ in., the length over the uprights 1 ft. 2 in., and the depth

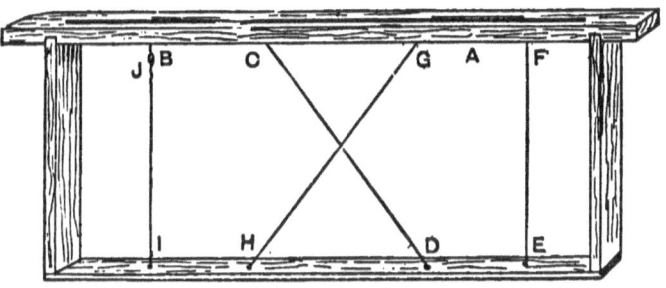

Fig. 39.—Wired Bar Frame.

over all 8$\frac{1}{2}$ in. The bottom and side bars vary in thickness with different manufacturers, but $\frac{1}{4}$ in. for the uprights and $\frac{3}{16}$ in. for the bottom bar, especially where the foundation is fixed by wiring, are to be preferred.

Materials for the bar frames, ready for nailing together, can be obtained from dealers, who also supply blocks for keeping the frames square while being nailed, but many bee keepers put the frames together without the blocks.

In working for extracted honey, shallow frames are now generally used. These have 1-ft. 5-in. top bars, and are 1 ft. 2 in. over the uprights, but the total depth is reduced to 5$\frac{1}{2}$ in.

52 *BEEHIVES AND BEE KEEPERS' APPLIANCES.*

However, the shoulderless frames in most general use are those of the American pattern, the joints in which are shown by Fig. 40; all these joints are accurately cut by machinery, and a frame block is not necessary for making them up. Frames cost so little that it does not pay to make them, except in quantity, and then only when a circular

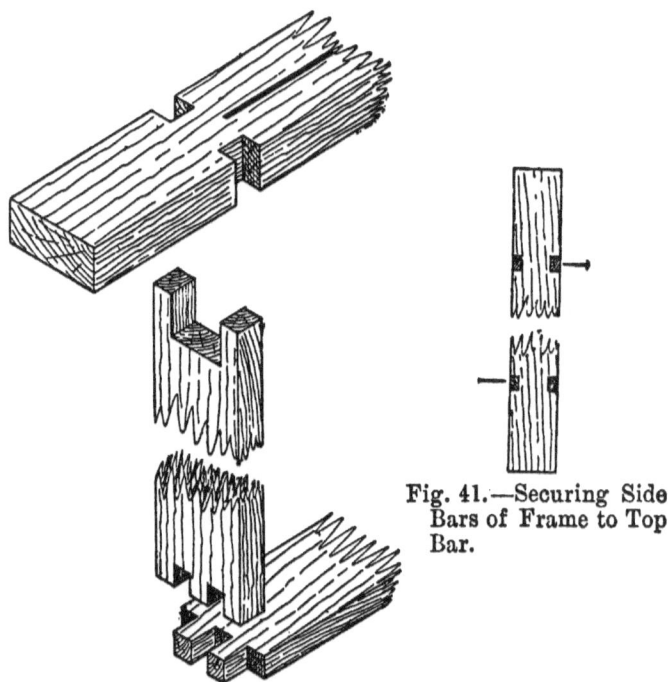

Fig. 41.—Securing Side Bars of Frame to Top Bar.

Fig. 40.—American Jointed Frame.

saw is available, and material that would otherwise be wasted can be used up.

These observations as to the home-making of frames apply with still greater force to broad-shouldered frames. Fig. 40 clearly shows how the parts of the frame fit together. Lay the top bar on the bench or table, force the tenons of the side bars into the mortises cut in the top bar, drive on the bottom bar, and after seeing that the side bars are square with each other, drive a fine wire nail through the tenon of the side bar into the top bar; one at each end on opposite sides will suffice (see Fig. 41).

Abbot's broad-shouldered frames are also made

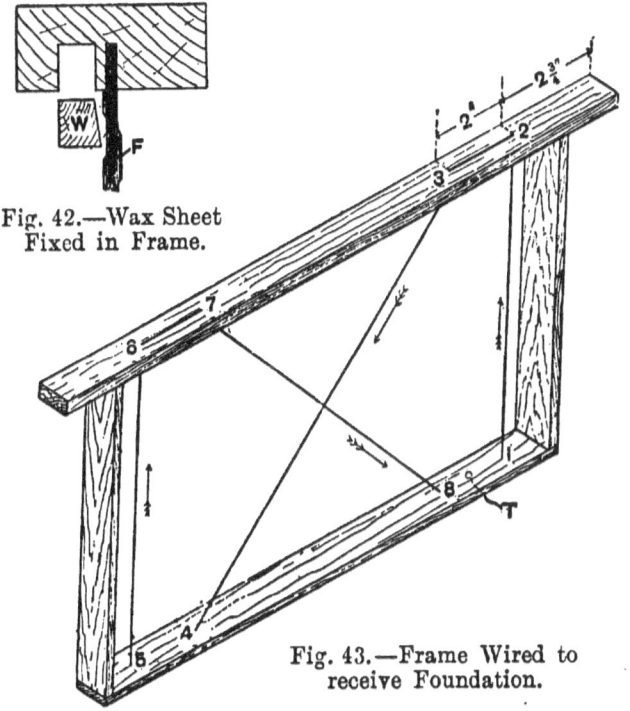

Fig. 42.—Wax Sheet Fixed in Frame.

Fig. 43.—Frame Wired to receive Foundation.

with mortise and tenon joints, but the mortises are cut centrally in the top bar.

To induce the bees to build combs in the bar frames instead of across them, it is necessary to use sheets of foundation wax, which are impressed with the bases of the comb cells. For the brood chamber these run about six or seven sheets to the pound, but thinner foundations are used in the sections, and by some in the shallow bar frames.

Some bee keepers use just a narrow strip of foundation fixed to the top bar of the frames, but it is far better and cheaper in the end to use a full sheet for each frame. When it is desired that the comb shall be wholly or in greater part built of cells "worker" size, it is advisable to use sheets of foundation large enough to fill almost entirely each frame.

Several methods of fixing foundations in frames are employed, the most common and the most objectionable plan being to insert the sheet of wax in a saw kerf cut in the centre of the top bar, as shown in Figs. 39 and 40. Reference was made to the weakness of the Association "standard" top

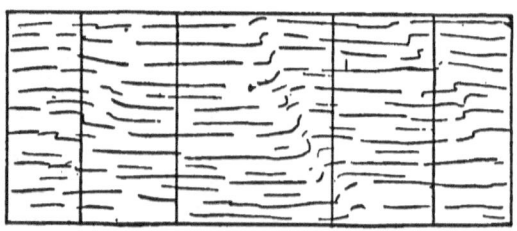

Fig. 44.—Block for Wiring Frame.

bar ($\frac{3}{8}$ in.); to put a saw kerf through it weakens it still further, and the cavity thus formed affords a hiding place for the larvæ of the wax moth—an insect which proves very destructive when it once gets possession of a comb. Fig. 42 represents a much preferable plan adopted by Abbot Bros. Two parallel grooves are cut on the underside of the top bar, the foundation F being placed in the narrower one; the wedge W is then driven tightly into the broader groove, by which means the wax sheet is very firmly secured.

When a solid top bar is employed, the sheet of wax may be fixed by running molten wax along at the junction of wood and foundation, and if this is done on each side of the sheet, and the wax is of the proper heat, a good joint will result. But a much more satisfactory plan of fixing full-sized sheets of foundation is to wire them as explained below.

A trellis of wire is secured to the frame, and these wires are heated and embedded into the wax itself so that the foundation cannot give way in the hive—an accident that no other method is proof against. Each apiarist usually has his own method of wiring frames; but that illustrated by Fig. 43 is general. Before making-up the frames, run a gauge mark down the centre of each top and bottom bar; then upon this gauge line, and at distances marked on Fig. 43, bore holes through with a fine

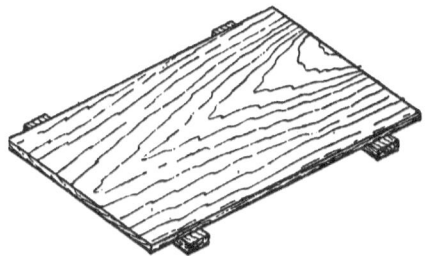

Fig. 45.—Gauge-board.

bradawl; also drive in a ¼-in. tack between one end pair of holes in the bottom of each bottom bar. After the frames are made up, thread some No. 30 tinned iron wire through the holes in rotation by the figures and in the direction indicated by the arrows, commencing at No. 1 and finishing at No. 8. Now fix the free end of the wire from hole No. 8 by giving it a twist or two round the tack at T, then, working backward, tighten the loop between 7 and 6, then between 5 and 4, 3 and 2, and finally pass the wire two or three times round the tack T, hammer the tack home, and cut off the surplus wire. If properly done, the wires, when touched, should emit a dull musical note, and neither top nor bottom bars should be strained out of parallelism.

Another method of wiring a frame is as follows: Prepare a block (Fig. 44) that will just fit loosely inside the frame (Fig. 44 is drawn to agree with Fig. 39). This may be about ⅜ in. thick, and should have four lines marked on it as shown. This block is dropped into the frame, and the positions where the lines intersect marked on the top and bottom bars. Holes are bored with a fine bradawl through

Fig. 46.—Woiblet Spur Embedder.

the top and bottom bars where marked in Fig. 39 at B, C, D, etc., those in the top bar being bored on the slant to miss the groove for the foundation. A length of wire is then taken from the reel, and a small loop B twisted at one end, and the opposite end passed up through the hole B, down through C and D, up through E and F, down through G and H, up through I, when it is put through the loop, pulled fairly tight, cut to length, and secured by twisting as shown at J.

For fixing the foundation to the wires, and also when the wax sheet is secured to the top bar by

Fig. 47.—Wheel of Woiblet Spur Embedder.

the smelting process, a guide-board (Fig. 45) is required. This is a piece of ⅜-in. wood, in size ⅛ in. less each way than the inside measurement of the frames, and to its underside two fillets are fixed crosswise, as shown, of a size so that they project at least ¼ in. on each side. Upon this guide-board lay a sheet of foundation, and above this lay the wired frame. Then a tool called an embedder

(Fig. 46) is heated sufficiently to melt beeswax, and run along over the wires, and as the wax melts the wires are pressed into the sheet, into which they are firmly embedded when the wax again cools.

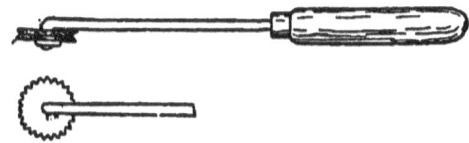

Fig. 48.—Spur Embedder with Wooden Handle.

The wheel of the embedder is heated to a black heat in a gas flame or fire; do not press the wire right through the wax.

The favourite tool for embedding is the Woiblet spur embedder, and Figs. 46 and 47 show in side elevation and in part plan, respectively, a home-made tool upon this principle. The wheel, cut from sheet brass, is ½ in. in diameter and ⅛ in. thick, and the wire handle is of ⅛-in. iron wire; a wooden handle instead of the wire loop would be an improvement (see Fig. 48).

A simpler embedder (Fig. 49) can be made in a

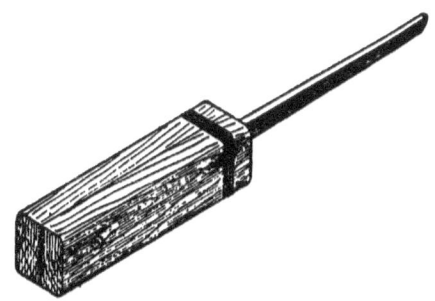

Fig. 49.—Embedder made with Floor Brad.

few minutes from a 3½-in. or 4-in. floor brad; two are useful, as one can be heating whilst the other is in use. An expert has had in use for several years a couple of these homely contrivances, which

58 *Beehives and Bee Keepers' Appliances.*

are as effective as more costly appliances. Take a floor brad (a cut nail will do), and placing it between two pieces of deal, about 4 in. long, 1 in. wide, and ½ in. thick, give the two a good squeeze in the bench vice to embed the nail in the wood. After removal, bind the nail end of the handle with two or three turns of copper wire and secure the wire ferrule with a touch of solder; then put a ¾-in. screw into the other end of the handle. Round off the point of the nail on one side, as shown by

Fig. 50.—Hoffman Self-spacing Frame.

Fig. 49, and after filing away the point until it is about $\frac{1}{16}$ in. thick, file a nick or **V**-groove in the rounded face, and the embedder is complete. The nick or groove is for the purpose of keeping the tool on the wire when in use.

The foundation having been fixed, and the frames being ready to be placed in the hive, the question of side spacing comes under consideration, that is, the keeping of the frames of comb the correct distance apart. As previously mentioned, frames are spaced laterally in the hive $1\frac{9}{20}$ in. (1½ in. bare) centre to centre, and whilst this spacing can be accomplished by marking the end of each frame at its centre, and placing these marks opposite similar marks made on the hive side, the heat of the hive is not conserved, as is the case when broad-shouldered frames or distance keepers are employed.

In America, and also, to some extent, in this country, a self-spacing frame, known as the "Hoffman," is very largely used; but in districts where propolis is abundant the bees glue the frames together to such an extent as to render a frame of this type a nuisance. Fig. 50 shows, in isometrical projection, a portion of two such frames. It may be said that the side bars are $1\frac{9}{20}$ in. wide at the

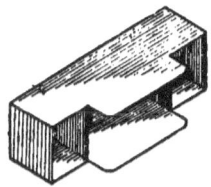

Fig. 51.—"W.B.C." Tinplate End.

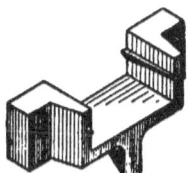

Fig. 52.—Cast Metal End.

top, and that the wide part on one side is reduced to a knife edge.

Broad-shouldered frames being self-spacing, the plain frames are the only ones that require the addition of metal distance keepers, technically known as "ends." Priority of place must be given to the "W.B.C." end, the invention of Mr. W. Broughton Carr. It is shown by Fig. 51, and is stamped out of a single piece of tinplate and pressed into shape by dies. Fig. 37 (p. 50) shows how these ends are used when the regulation $1\frac{9}{20}$-in. spacing is adopted; Fig. 38 (p. 50) shows how the ends are placed on the frames when $1\frac{1}{4}$-in. spacing is required; Fig. 52 shows a cast metal end somewhat largely used; Fig. 53 shows the "Howard" end, made of tinplate; and Fig. 54 shows Dr. Pine's cast metal end. There are probably more "W.B.C." ends used than all the others put together; they are made both for ⅜-in. and ½-in. top bars, and a special wide end is also made—for ⅜-in. top bars only—for fitting to extracting combs, eight of the wide ends occupying the space of ten of the ordinary pattern. They can be obtained in two widths, 1½ in. bare for the brood chamber and 1¾ in. for the shallow-frame supers. When using 1¾-in. ends in the supers it is a good plan to use "drone" foundation.

Whilst upon the subject of spacing frames, mention may also be made of end spacing. If hives are made according to instructions given in

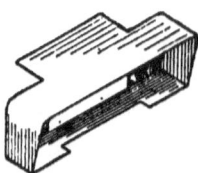

Fig. 53.—Howard Tinplate End.

Fig. 54.—Pine's Cast Metal End.

Chapters II. and III., end-spacing appliances are not necessary, as the top bars being cut accurately to length, make a perfect fit with the hives, and no end-shake will be possible. But in roughly made hives some means to counteract any inaccuracy in construction will be necessary, and it will be noticed in Figs. 52 and 54 that each end also has an angular spur on the underside, the spur allowing the regulation ¼-in. spacing between the hive wall and the frame end.

For use with the ends illustrated by Figs. 51 and 53, and also with the Hoffman self-spacing frames (Fig. 50), a method of end spacing is shown in Fig. 55, a small staple being driven in each side bar under the lug so as to project ¼ in. The same plan is adopted for bottom spacing, as shown in the same figure, this provision being made for the purpose of ensuring that bees—the queen especially—shall not be injured by being squeezed between the hive side and the frame when the combs are being withdrawn or replaced.

As bees build their combs about an inch in thickness, there is, nominally, a space of ½ in. between the faces of capped brood on two adjacent combs. As the outside face of each outside comb only gets, under ordinary circumstances, half of this working space, two plain strips of wood (see P S, Fig. 17,

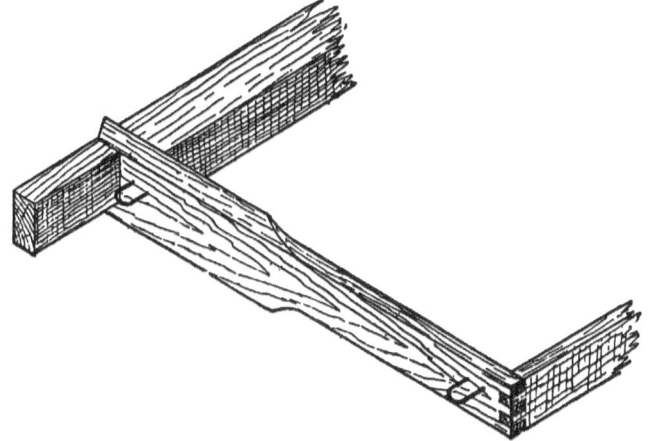

Fig. 55.—Staples used as End Spacers.

p. 31) ⅜ in. by ¼ in. in section are used in order to increase the distance between each outer comb and the hive side or the "dummy" (D S, same reference).

A dummy is a cleated or clamped board which fits the inside of the hive, and has a top bar, by means of which it can be suspended in the same way as a frame. Its use is to contract the brood nest when it is desired to give the bees a lesser number of frames than the hive provides accommodation for; and in the hive described in Chapter III. a pair of cleated dummies is used to make the hive double-walled at the sides when ten or a lesser number of frames are in use. The sizes figured on Fig. 56 are for the above-mentioned hive, although dummies of this size will be equally serviceable in any other hive made to take "standard" frames, as long as the thickness is not greater than accommodation can be found for. Fig. 57 shows

Fig. 56.—Back View of Dummy.

this dummy in part section, and by the dimensions given in Figs. 56 and 57 it will be seen that its total thickness is 1 in.; consequently ten frames (14½ in.), two spacing slips (½ in.), and two dummies (2 in.) exactly occupy the space provided in the hive described in Chapter III.—namely 17 in. The dummy should be an easy fit inside the hive, and its dimensions should be 14½ in. (bare) long and 8½ in. (bare) deep under the top bar. The illustrations supply other measurements.

A still simpler dummy is shown by Fig. 58. This is a solid frame, and (to fit the hive described in Chapter I.) may be cut from a piece of ⅝-in. pine,

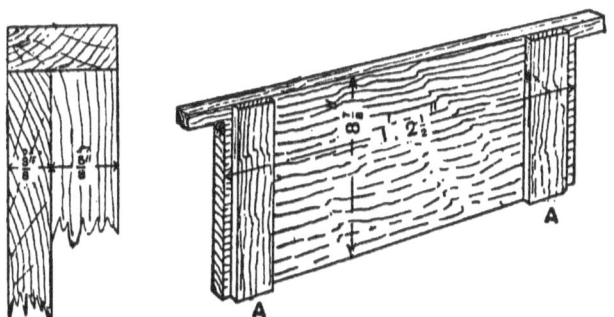

Fig. 57.—Section of Dummy.   Fig. 58.—Simple Form of Dummy.

1 ft. 5 in. long by 8⅞ in. wide, if it is to prevent bees getting behind the dummy, although some bee keepers prefer to have the dummy with a ⅜-in. space underneath, to allow the bees to return to the nest if they should get into the top or behind the dummy when the hive is examined, and for this reason make it 8½ in. wide. It is a good plan either to

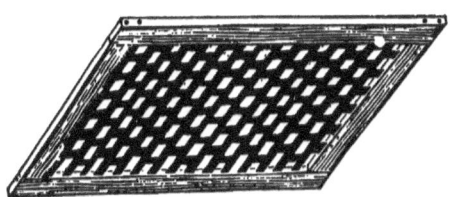

Fig. 59.—Old Style of Queen Excluder.

clamp the ends, or to secure ledges A across the grain of the dummy to prevent warping.

To prevent the queen bee getting up into the sections, excluder zinc, which is pierced with slots through which only the workers can pass, is used over the brood nest. This can be obtained in sheets 16 in. square, and is best laid over the frames without being fitted into a light wooden frame as used to be the custom. (This old style is shown by Fig. 59.)

Quilts are necessary for covering the frames,

Fig. 60.—Section with Foundation.

and for this nothing is better than a square piece of American cloth of good thickness laid on with the painted side downwards. Sheets of celluloid can be obtained for the purpose, through which the bees can be seen at work; they are very good, but rather expensive. Over the first quilt a number of thicknesses of carpet or sacking of some kind should be placed to keep up the warmth of the hive, and for this woollen material is the best, but it harbours moths, and for this reason sacking or cotton material is often preferred.

Fig. 61.—Section before Folding.

For marketing honey in the comb, sections (Fig. 60) are used. These are wooden boxes, generally 4¼ in. by 4¼ in. by 2 in., made of thin bass-wood, and they each hold just a pound of comb honey. Larger sizes can be obtained. The wood for making them ready prepared can be bought in gross

FURNISHING AND STOCKING A BEEHIVE. 65

lots for about 2s. 6d. The pieces are V-jointed, as shown by Fig. 61, and the ends are machined to fit together when the box is folded over. The top side has a saw kerf run in it, which is used for the insertion of the comb foundation, a triangular piece of which is generally employed as shown in Fig. 60.

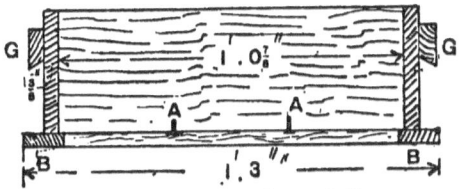

Fig. 62.—Cross Section of Crate.

If the wood is very dry, the joints should be damped before folding the sections together, and twenty-one of these sections are packed into a crate, which is shown in section by Figs. 62 and 63.

To make a crate (see Figs. 62 and 63), two pieces of ⅜-in. deal 1 ft. 0⅞ in. by 4½ in., and two 1 ft. 3¾ in. by 4¼ in., are nailed together to form a bottomless box. This is divided into three equal spaces, lengthways, by metal bars (Fig. 64) placed as shown at A (Fig. 62). These can be obtained from dealers,

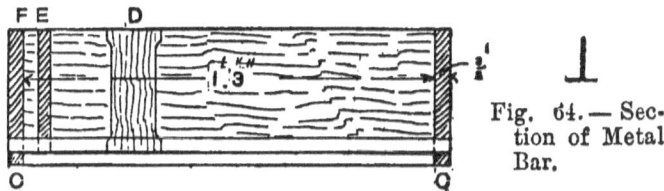

Fig. 64.—Section of Metal Bar.

Fig. 63.—Longitudinal Section of Crate.

and are fixed by driving into a fine saw kerf. Two strips B, each 1 ft. 3¾ in. by 1¼ in. by ⅜ in., are then nailed on the bottom of the crate lengthways, and two pieces C (Fig. 63), 1 ft. 0⅞ in. by ⅜ in. by ⅜ in., between them across the ends. The crate is then filled with the sections, one of which is shown in

E

position at D, but if good results are expected, zinc or wood dividers (Fig. 65) should be placed between the sections, six being used in each crate.

When the sections are placed in the crate, a follower E (Fig. 63) is required to keep them in position. This consists of a piece of ⅜-in. deal 1 ft. 0¾ in. long by 4¼ in. wide, placed in the position shown and forced against the sections by means of a spring, slipped in at F, or a cork cut to length will serve the purpose. A couple of blocks G (Fig. 62) are nailed on at the ends for lifting the crate off the hive.

The following information on stocking a hive is of general application, but also refers particu-

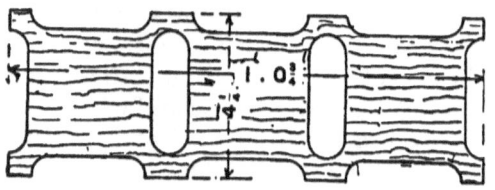

Fig. 65.—Section Divider.

larly to the simple hive described in Chapter I., and illustrated by Figs. 1 and 2 (pp. 12 and 13):—

First fill the lower portion of the hive, or body box, with full sheets of brood foundation wired into the frames. Eight or nine frames will be required if stocking with driven bees during the autumn. About 5 lb. of driven bees, costing about 6s. 6d., will be necessary. If possible, obtain the bees from a dealer in direct railway communication, or delay or knocking about will have killed them before they arrive. Shake them out of the box on to the frames, and cover with a piece of American cloth, in the centre of which a hole is cut.

Previous to the arrival of the bees, have prepared a quantity of feeding syrup made by boiling 4 lb. of sugar in a quart of water, to which a dessert spoonful of salt and the same quantity of vinegar have been added. Stir the syrup whilst on the fire, and do not let it burn. Boil for three or four minutes. Place the syrup in a large glass jar, tie a piece of open textured calico over the mouth, and invert it over the hole in the quilt on the frames, when the bees will suck it down. When the jar is empty, slide a piece of tin underneath, or smoke the bees down, and take it off, refill, and replace. From 20 lb. to 25 lb. of syrup must be fed to the bees to enable them to live through the winter, and all feeding must be completed by the end of September. When the feeding is finished, remove the frames not covered by bees, close up the dummy to the side of the frames, and cover up warmly till the spring.

Another method of stocking is to buy in the spring a straw hive containing a good stock of bees ; place this on the top of the frames, of which ten should be used, all filled with full sheets of foundation. Close all openings at the top to make the bees work through the proper hive entrance, and by the autumn the bees will have transferred the breeding place to the frames, and the straw skep filled with honey can be removed. Still another method is to purchase a strong swarm of bees in May, but in this case honey is seldom taken the first year.

CHAPTER VI.

OBSERVATORY BEEHIVE FOR PERMANENT USE.

There are two kinds of observatory beehives. The hive seen at flower shows, in which from one to three frames of comb with bees are placed for a few days on exhibition, is not suitable as a permanent home for the bees, as sufficient heat cannot be maintained by them to hatch eggs and rear brood except under very favourable conditions. This form of hive is fully described in the next chapter.

The observatory hive forming the subject of this chapter is a permanent home for the bees, special provision being made for wintering, and tiering up in the summer for gathering surplus honey. Ten standard frames are used in the brood chamber, and at the sides are flaps which may be opened so that the bees may be observed through glass frames in the inner shell of the hive. The outer flaps must be kept closed except when the bees are examined, or, having great objections to light, they will cover the glass with propolis; it is a good plan to place a cushion of some thick material between the flaps and the glass, especially in winter.

To make the observatory hive, first prepare pieces of red deal for the stand. Two side-pieces A (Fig. 66), 2 ft. 3 in. long by 2½ in. by 1¼ in., and two cross-pieces B and C, 1 ft. 5½ in. long by 2½ in. by 1¼ in., will be required. Bevel off the front for the alighting board as shown, and mortise the cross-pieces into the sides.

For the floor-board, prepare a piece of pine 1 ft. 6½ in. by 1 ft. 6¼ in. by ⅞ in. thick, and in this sink a recess 12 in. wide by ⅜ in. deep and sloping up

OBSERVATORY BEEHIVE FOR PERMANENT USE. 69

to about the centre of the hive as shown at D. The floor-board may be nailed to the cross and side pieces. Nail the alighting-board E to the sloping

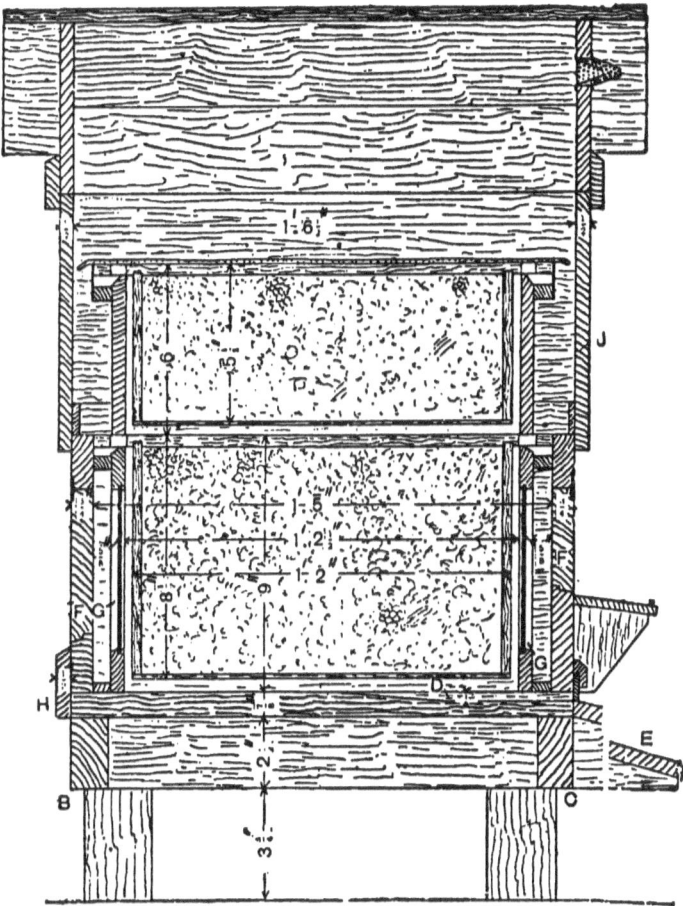

Fig. 66.—Vertical Section through Observatory Beehive.

front of the side-pieces, taking care that the top comes level with the bottom of the recess in the floor-board.

The brood chamber is a plain bottomless box of ⅞-in. deal, 1 ft. 5 in. by 1 ft. 4¾ in. inside and

70 BEEHIVES AND BEE KEEPERS' APPLIANCES.

barely 9 in. deep. On each of its four sides a square hole is cut as F for the reception of the inspection flaps, a bevelled rebate being prepared at

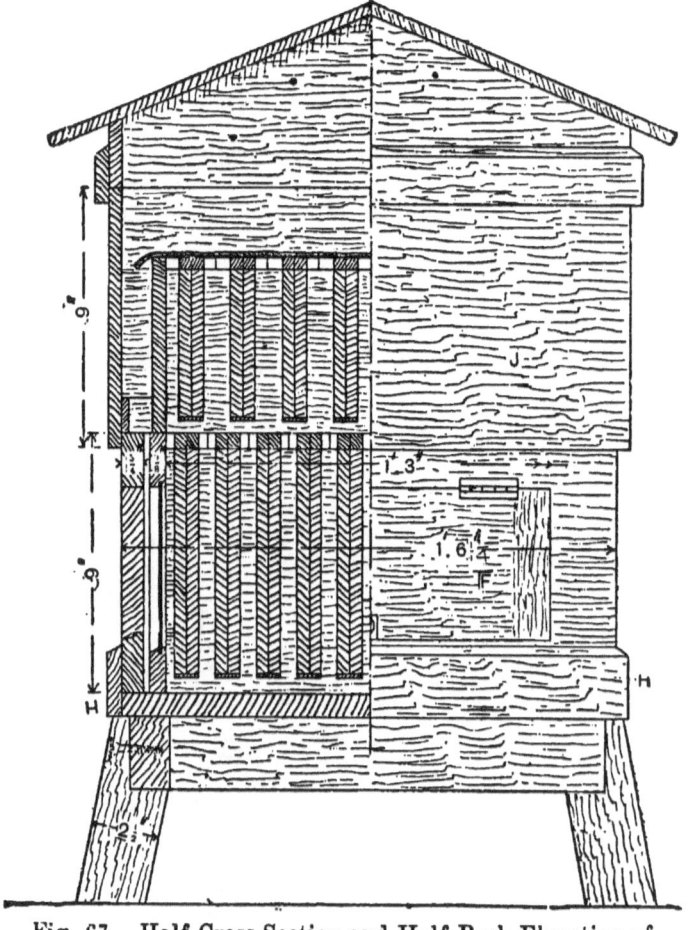

Fig. 67.—Half Cross Section and Half Back Elevation of Observatory Beehive.

the bottom of each to form a stop and keep out the wet, as shown in Figs. 66 to 68. Across the front and back two frames G (Fig. 66) with glass are fixed so as to leave a distance of 1 ft. 2½ in.

between them. The frames should be fixed flush with the bottom of the box and ⅜ in. below the top. Fillets should be nailed between these and the outer

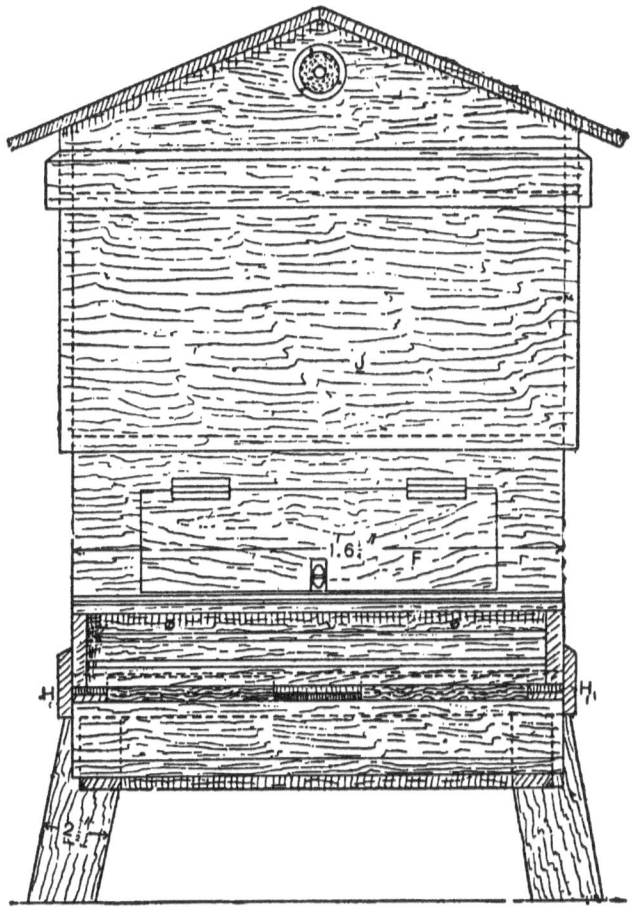

Fig. 68.—Front Elevation of Observatory Beehive.

case, the bottom ones being flush and the top ones ¾ in. down (see Fig. 66). On the bottom of the back and sides a 2½-in. by ½-in. plinth H (Figs. 66 to 68) is nailed to keep the wet from getting under the floor.

A porch at the front of the hive to protect the bees when returning home in wet weather, and sliding shutters to regulate the size of the opening, the construction of which will be clear from Figs. 66 and 68, are shown. The porch should be fixed with screws so that it can be lowered in the winter and raised in the summer. The side flaps F should be clamped and attached by means of brass butt-hinges, and fastened at the bottom with small flush bolts; or turn-buttons may be used if preferred.

Over the brood chamber a lift J is shown. This consists of a bottomless box made of 9-in. by ½-in. stuff, the inside dimensions being equal to the outside of the brood nest. The sides of this and the brood nest may be dovetailed together or lapped, or butt-jointed and nailed. It is a good plan to paint well all joints before putting them together. Three-eighths of an inch from the bottom of the lift a 1-in. by ¼-in. strip is nailed on all round (see Fig. 66). In the summer the lift is used as shown in the illustrations, but in the winter it may be turned upside down to cover the flaps and keep out the wet. A further lift may be used in the summer if required for working for extracted honey, but it should be made with a plinth the same as the roof.

The construction of the roof will be clear on reference to the illustrations. A pair of cone bee-escapes are fitted in the front to allow any bees that may get from under the quilt to escape. It is advisable to cover the roof with sheet zinc.

The interior fittings are shown in the sections (Figs. 66 and 67). In the brood chamber are ten frames of comb with W.B.C. ends. On each side of the set of frames dummies filled with glass are used, so that when the side flaps are open the combs can be seen. A section of one of the dummies is shown at the left of the frames in Fig. 67. In the winter two or more of the frames may be taken out and the dummies moved closer to the centre of the hive and the space outside filled with some warm material.

The super over the brood chamber is made to take eight broad-shouldered frames or ten ordinary shallow frames. It consists of a box 14½ in. by 15 in. by 6 in. deep, of ½-in. deal. The ends are ⅜ in. shallower than the sides at the top (Fig. 66), to allow for the shoulders of the frames, which are kept in position by a strip of wood nailed on outside, on which is tacked a piece of tin brought up flush with the top of the frames.

For the quilt covering the frames a piece of American cloth should be used with the glazed side downwards, or a transparent celluloid quilt through which the bees can be seen at work may be used if preferred. Over this a piece of carpet or any thick, warm material such as sacking should be placed. Woollen material of any kind is not recommended for quilts, as it harbours moths, the warmth of the hive proving very attractive to these insects (see p. 64).

CHAPTER VII.

OBSERVATORY BEEHIVE FOR TEMPORARY USE.

The leading features of unicomb or observatory hives for temporary use are the same, the differences being in the details of construction and mode of setting up. Indeed, all that is required is a case glazed on opposite sides, within which a single cake of comb can be suspended in such a manner as to allow no greater distance from the glass than is sufficient to allow the bees to cover the comb on both sides. This distance may be taken as about $2\frac{1}{8}$ in.; a little less might be allowed without inconvenience, provided selected combs, perfectly flat and even in thickness, are used.

The width and depth of the case will depend on requirements, and as it is best to stock an observatory hive with frames containing comb from a colony already established in a bar-frame hive, the dimensions must be adapted to the size and number of frames it is proposed to use. The smallest size will be that suited to a single standard frame 14 in. by $8\frac{1}{2}$ in., but it is more common to have these observatory hives to hold two, three, four, or six frames.

The observatory hive, illustrated by Fig. 69, is to hold three frames of the size above mentioned, but to show its construction better, only the central frame is drawn in position. Fig. 70 is a vertical cross section, Fig. 71 a horizontal section, and Fig 72 a sectional plan of the top.

Mahogany or pine is a suitable wood for making the hive, or both may be used. The internal frame A (Fig. 69) chiefly determines the sizes of the other parts of the hive, and for this reason it should be

OBSERVATORY BEEHIVE FOR TEMPORARY USE. 75

made first. This frame is shown in detail, with dimensions, by Fig. 73. The internal dimensions should be exactly followed, but the frame should

Fig. 69.—Observatory Beehive for Temporary Use.

be left a trifle full outside in order to facilitate the fitting to the case into which it is inserted, by taking shavings from each of the sides until an exact fit is obtained.

76 *Beehives and Bee Keepers' Appliances.*

The thickness of the frame should be 1⅜ in. if the glass windows are fitted as shown by B (Figs. 69 to 71); but if the glass is to be fixed by a bead or strips of wood against the frame A, an increased

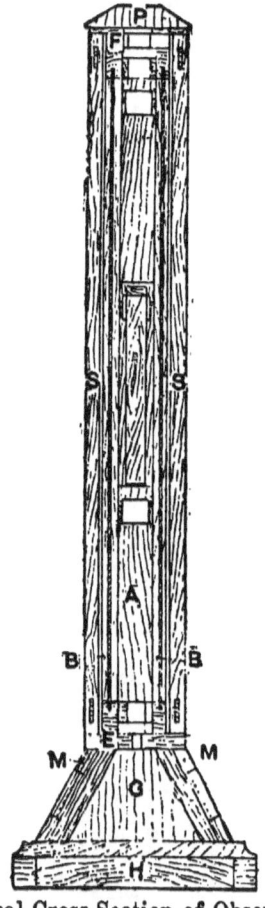

Fig. 70.—Vertical Cross Section of Observatory Beehive.

thickness will be necessary. The top and bottom pieces of the inner frame are fastened at the corners by screws or nails as shown in Fig. 73. The notches in the side pieces are cut with a tenon saw and a chisel to a depth of 1¼ in., just sufficient to

OBSERVATORY BEEHIVE FOR TEMPORARY USE. 77

admit freely the top bar of the frames. The projecting ends of the top bar rest on the edges of small pieces of tinplate as shown in Fig. 69. If it is intended to fix the inner frame to its outer case by means of dowels through the sides, the holes

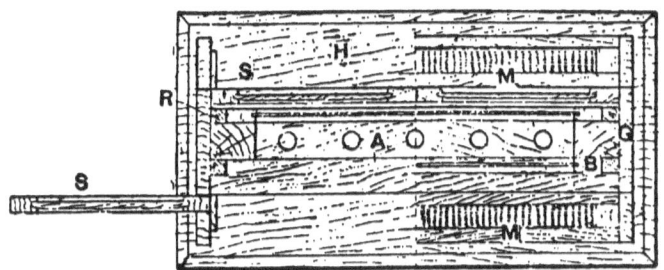

Fig. 71.—Horizontal Section of Observatory Beehive.

for them should be bored before the inner frame and case are jointed up.

It will be seen from Figs. 69 to 71 that the side pieces of the case fit close to the sides of the inner frame, but the top and bottom pieces F and E are clear by a space of about ⅜ in. and ½ in. respectively. The shape of the side pieces of the case G is seen in Fig. 69. The top piece F is fixed to the sides by dovetailing, or by nails or screws, and the bottom piece E slips into a groove cut in the side pieces.

The base H (Figs. 69 to 71), into which the lower

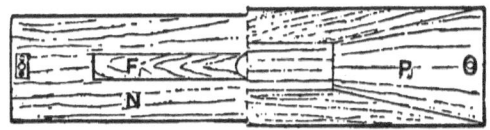

Fig. 72.—Sectional Plan of Observatory Beehive Top.

ends of the side pieces are slightly sunk, is a flat board with a moulded or chamfered edge, supported on four edging pieces mitred together at the corners. The piece E is perforated with a number of ½-in. holes, which form a communication with

the interior of the lower part of the case. This is
closed in on both sides with frames M fitted with
panels of perforated zinc or wire cloth. This lower
chamber is intended as a means of giving ventilation to the hive when the bees are confined by
closing the entrance L (Fig. 69), when no provision
for flight can be made. At other times the holes
in the piece E are closed by sliding a thin slip of
wood or vulcanite through the doorway L between
the bottom of the inner frame A and the top of the
piece E, thus cutting off the communication between
the hive and the ventilating chamber below.

Above the top piece F, and screwed to it, is a
piece of board $\frac{1}{2}$ in. thick, with a rectangular slot
cut in its central part. This piece N (Figs. 69 and
70) is shown in half-plan on the left of Fig. 72, and
it will be seen that its breadth is the same as that
of the pieces forming the case, and that it projects
about 1 in. at each side. Below these projecting
ends, and against the sides G (Fig. 69), blocks o, to
strengthen the corners of the case, are fastened by
screws or dowelled and glued.

Surmounting the case is a cap, P, the upper edges
of which are bevelled; it is secured to the case with
thumbscrews, so that it may be easily removed for
bottle-feeding the bees through a hole in P (see
Fig. 72). Within the slot in N a piece of perforated
zinc should be loosely fitted to keep the bees from
coming through the holes. If it is not intended to
use these holes as ventilators when the bees are
confined to the hive, any simple means may be used
to cover up the holes.

If desired, a bow handle by which to lift the
hive can be fastened to the flat part of the cap,
in which case the small plates into which the
thumbscrews hold must be fixed to the underside
of the piece N and sunk, and not as shown in the
left-hand part of Fig. 72.

The two windows are frames made up of four
pieces with a groove ploughed in one edge and

OBSERVATORY BEEHIVE FOR TEMPORARY USE. 79

mitred together at the corners. Half a window is shown in the right-hand part of Fig. 69. The inside measurements of the window frames are the same as those of the inner frame A, and the breadth of

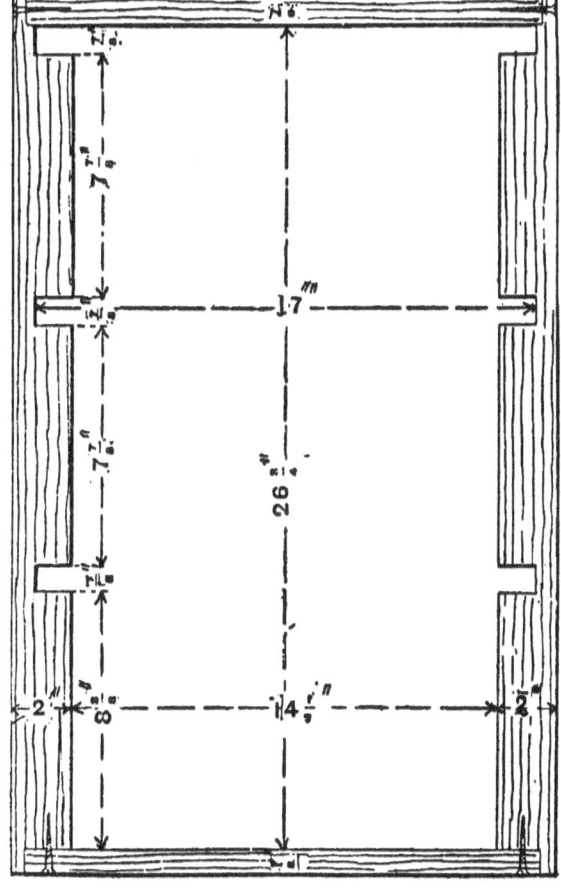

Fig. 73.—Inside Frame of Observatory Beehive.

the top and side pieces is 1¼ in. The bottom piece is 1⅜ in. wide, which throws the line of the bottom corner joints a little out of a true mitre of 45°. This can be obviated by making the four pieces of the window frame 1¾ in. broad, a small reduction in its internal breadth and depth being allowed to suit. On each side of the windows, and fixed to the case, are the strips R (Figs. 69 and 71), to which small catches, that secure the windows in their places, are screwed. Four shutters S (see also Fig. 70), hinged to the case, are provided to shut out the light as well as to retain heat. They are framed and filled in with a panel, and should be provided with a bolt at the top and bottom of one of the shutters, and a latch on the other. Some bee keepers recommend lining the inside of the shutters with felt or other heat-retaining material; but as this hive is only for use in the warm season of the year, this is unnecessary.

The hive should be set up within doors; a warm out-house or shed will do very well, or a room in a dwelling-house might be used for the purpose. A covered passage must be made through the wall to the outside for the bees. A small bracket landing place for the bees is shown by dotted lines close to the entrance L (Fig. 69).

In stocking an observatory hive of this kind, select three combs from the centre of a strong colony in a bar-frame hive containing plenty of brood in all stages from the egg upwards. Lift them out, and place one at a time into the unicomb hive with the bees clinging to them, taking care that the queen bee is also moved in. It is advisable that a large proportion of the bees to populate the observatory hive should be young ones that have not flown, otherwise there is great risk of depleting it, through the older bees going back to the hive from which they were taken. On this account it is preferable to operate on a fine, warm day, at a time when many of the older bees are out, and it is then a comparatively easy matter to take with a stiff feather or piece of card from several frames as many young bees (known by their lighter colour) as are required.

OBSERVATORY BEEHIVE FOR TEMPORARY USE. 81

It is quite possible to succeed without taking the queen with the bees, if the combs contain

Fig. 74.—Mounting Observatory Beehive on Brackets with Pivots.

eggs, or larvæ not more than three days old, from which the bees can raise a young queen, but in

F

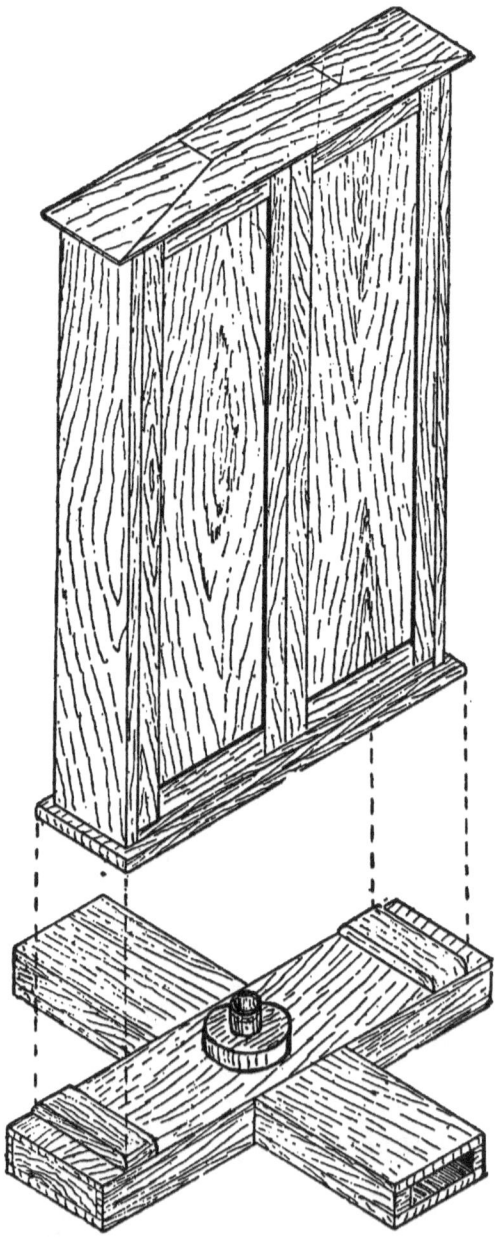

Fig. 75.—Mounting Observatory Beehive on Feet with Pivot.

this case considerable time elapses (about six weeks) before the progeny of the young queen is added to the population, which, meanwhile, will diminish somewhat in number, so that results may not prove as satisfactory as by the adoption of the first method.

There are various ways of mounting observatory hives besides that illustrated in Fig. 69. Fig. 74 shows in section a hive swung between pivots, so that it can be turned with either side to the wall. The lower pivot is of wood with a central hole through which the bees pass to the hive, a hole being cut through the supporting block up which the bees may creep. This block comes against the opening cut in the external wall of the room against which the board, to which the whole is fixed, is bolted. The top pivot can be withdrawn so that the hive case may be detached.

The hive can also be supported on a central pivot as shown in Fig. 75. The foot is cross-shaped, one of the pieces being hollow and forming the entrance for the bees. In the interior of the foot immediately below the hollow pivot an incline of wood is placed to guide the bees to the hole that leads into the hive.

CHAPTER VIII.

INSPECTION CASE FOR BEEHIVES.

An observatory hive is an expensive luxury that few amateur bee keepers can afford; and although an inspection case (Fig. 76) will not take its place for exhibition purposes it will be found useful

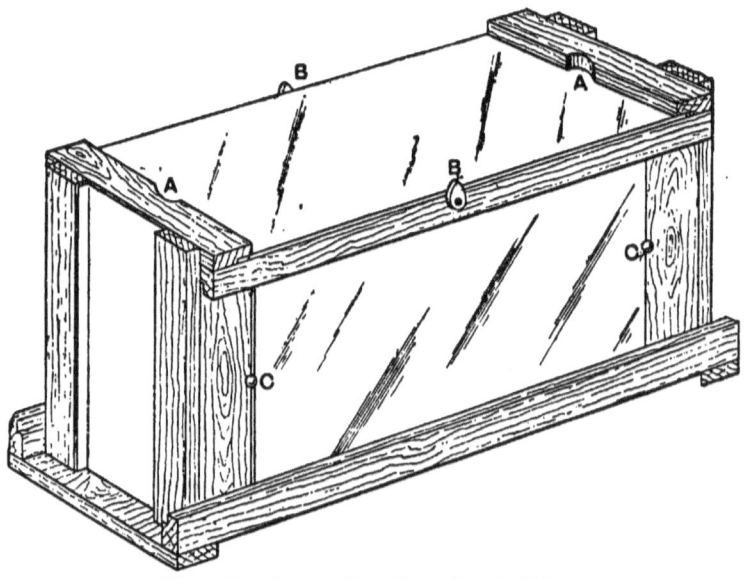

Fig. 76.—Inspection Case for Beehives.

when showing the working of a hive to timid persons, or for examining the frames on a cold day when there is danger of the brood being chilled, or the queen blown off the comb. It will also be found useful to those bee keepers who strongly object to being stung. The frames may be lifted one after another into the case and examined on both sides; a frame may be transferred to another hive; queen cells may be cut out by sliding back the glass slightly to insert a thin-bladed knife

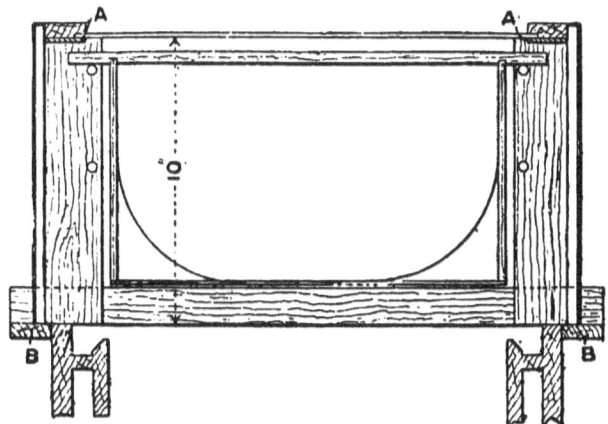

Fig. 77.—Section of Inspection Case.

for that purpose; or a frame of honey may be removed from the hive by lifting it into the case and driving out the bees with smoke or carbolic, when such a course happens to be necessary.

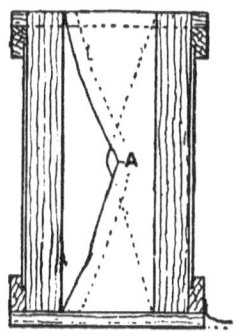

Fig. 78.—End Elevation of Inspection Case.

The case is made of ½-in. pine throughout, of a length to take standard size bar-frames. The total length must be determined by the size of the hives for which the case is made. Fig 77 shows the case on a hive, with the outer casing, beyond the end of the bar-frames, ¾ in. thick. Fig. 78 is an end view of the case. If two sizes of hive are used, the case may be made to the longest hive and a piece of calico tacked on from side to side, as shown at C (Fig. 79).

To make the case, proceed as follows: Cut two pieces of pine, 1¼ in. by ½ in., and 3 in. longer than the length of hive, and two pieces, 1 in. by ½ in., equal to the length of the hive. Plane these up and rebate them ¼ in. deep in the width, and the depth of the thickness of a piece of window

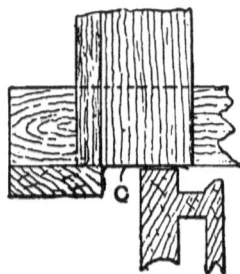

Fig. 79.—Arrangement of Case to Fit Two Lengths of Beehives.

glass (15 oz.) in the thickness. Next get out four pieces 10 in. long by 2 in. by ½ in.; nail the long pieces to the short, as shown in the drawings, and this will make the two sides.

Now prepare the top, bottom, and end pieces. For the top, two pieces will be required 7 in. long by 1½ in. wide by ½ in. thick. Rebate these for a piece of window glass to form the top, and cut a piece out as at A (Fig. 76) to allow of the frames being lifted out if required. The size of pieces cut out must be regulated by the length of the frames, and the rebate worked to the full depth of the pieces cut out. Two light pieces are nailed on the bottom of these top pieces, as shown at A

INSPECTION CASE FOR BEEHIVES.

(Fig. 77), to form a groove to prevent the glass falling into the case.

The bottom pieces are 7 in. long by 1½ in. by ½ in., and are to be nailed on the projecting ends of the bottom side pieces, to form a rebate to keep the case in position on the hive.

Four end pieces are required, 10½ in. by 1¼ in.

Fig. 80.—Securing End Openings of Inspection Case.

by ½ in., and are nailed to the top pieces, sides, and through the bottom pieces, which will complete the woodwork.

Three pieces of glass will next be required, one for the top, 1 ft. 5 in. long by 6 in. wide, and two to fill the spaces at the sides. Two little buttons are screwed on, as shown at B (Fig. 76), to keep the top glass in position, while four small screws c

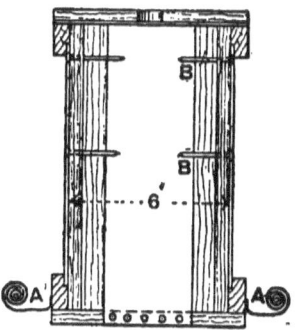

Fig. 81.—Cross Section of Inspection Case.

will serve to keep the side glasses from falling off. The spaces at each end are filled with four pieces of calico or other suitable material, and an arrangement is made here to get the fingers in to lift the frames without letting the bees get out. Fig. 80 will explain how this is done. One edge of each piece of calico is hemmed, and a piece of elastic drawn through A. The calico is tacked on at the bottom, top, and sides, leaving the elastic edges free. A (Fig. 78) shows the elastic edges drawn back ready for the insertion of the fingers. Two pieces of calico will now be required, each large enough to cover the whole top of the hive. These are tacked to the bottom bars of the case, as shown in Fig. 81. To complete the case, eight 2-in. wire nails are driven through the sides at

Fig. 82.—Handle for Lifting Frames.

B (Fig. 81), on which to hang the frames for examination.

When using the case, take off the cover of the hive; begin from one side, and, as the quilt is rolled off, push the case on. The frames may then be lifted up into the case, as shown at Fig. 77, and hung on the nails. A couple of hooks (Fig. 82), made with a piece of steel wire and a couple of bradawl handles, may be used for lifting the frames.

CHAPTER IX.

HIVE FOR REARING QUEEN BEES.

BEE keeping, conducted upon advanced principles, requires the bee keeper to provide himself with a supply of fertile queen bees during the working season. These are introduced into stocks that have been swarmed by art, in order to supply the place of the queen taken from them; or, in the case of natural swarms, to save the time—very precious during the honey-flow—that would elapse before the immature queens left on the departure of the first swarm arrive at maturity, and take up the maternal duty of keeping up the working population of the hive, or of the new colonies which the secondary swarm or swarms originate. They will also be needed to enable the bee keeper to carry out certain other objects he may contemplate, such as the supersession of queens whose powers are declining through age, or which lack the desirable qualities good queens possess.

The rearing of queens is then an important matter with bee keepers who have apiaries of more than a few stocks of bees. The work is usually done by setting apart several small colonies of bees obtained by the division of full-sized stocks. The queens are reserved, so that the colonies are queenless, and in each of these queenless colonies a selected queen cell eight or nine days old is placed. A few days after the queen will arrive at maturity and liberate herself from the cell, and at a later stage, if all goes well, she will be found busily employed laying eggs, and receiving the attentions of her subjects. As soon as the bee keeper is assured of the queen's fertility by the presence of eggs, or better, of worker brood in the hive, he may utilise her as he thinks proper.

Full-sized hives, with the interior space contracted by means of division-boards, so as just to take in three bar frames, serve for queen-rearing, but it is generally more expedient to use special or nucleus hives of simple construction and smaller size, which are easy to manipulate or move about.

The hive, of which a perspective view is given in Fig. 83, is half the size of the usual full-sized

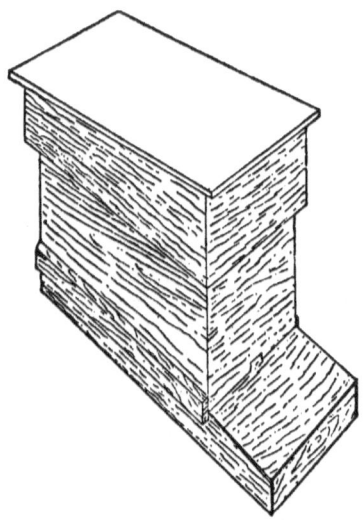

Fig. 83.—Hive for Rearing Queen Bees.

hive. It takes in five standard size bar-frames, and will serve either for queen-rearing or to accommodate temporarily a small or medium swarm. It is not difficult to make, and to a beginner in bee keeping who proposes to make his own hives it will afford a preliminary exercise before taking in hand the construction of stock-hives.

A plan of the body box, and as much of the base-board as can be seen by looking directly downwards, is shown in full outline in Fig. 84; and Fig. 85 is a longitudinal section taken centrally through the roof, body, and base, or footboard.

As the hive is to hold standard frames, the size of the frame determines the dimensions of the

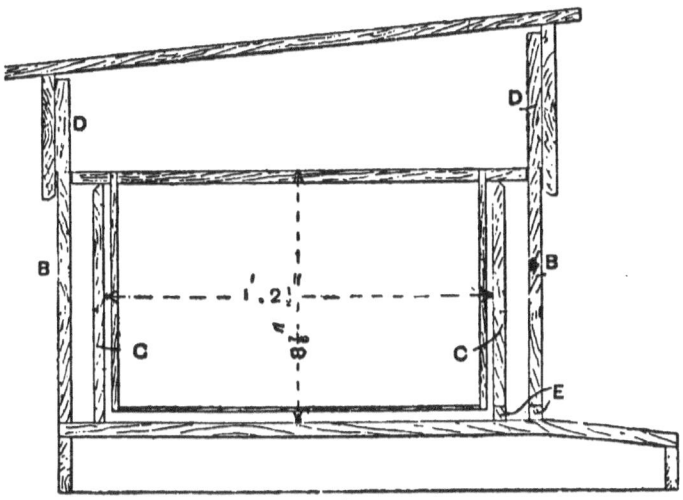

Fig. 84.—Plan of Body-box.

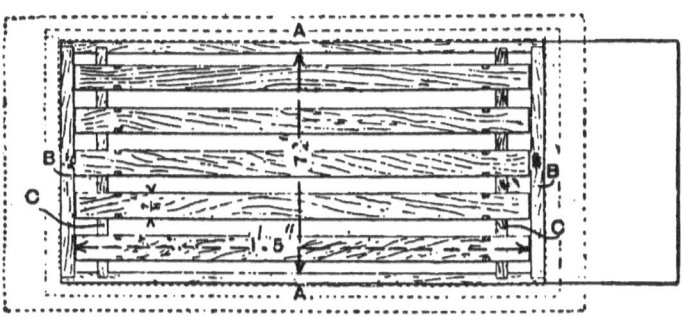

Fig. 85.—Section of Hive for Rearing Queen Bees.

body box. The outside measure of the standard frame is 14 in. by 8½ in. by ⅞ in. The top bar is 17 in. long and ⅜ in. thick; the side pieces and the bottom bar are ¼ in. and ⅛ in. thick respectively. The space between the frame ends and the inside of the hive body is ¼ in., so that the inside length of the body box is 14½ in. The inside width, to take in five frames, each 1½ in. from centre to centre, is 7¾ in., and the depth to give a ⅜-in. space below the frames is 8⅞ in. These three regulating dimensions are marked on Figs. 84 and 85, and as the illustrations are drawn to scale, other measurements may be taken from them.

The body of the hive is made up of six pieces: two sides A, two ends B, and two inside ends C.

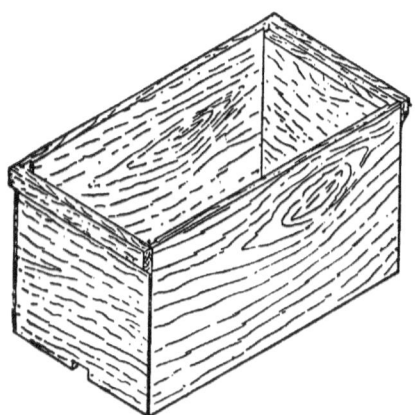

Fig. 86.—Modified Body-box.

The ends B fit into rebates in the sides A, and the inside pieces C into grooves cut in the sides, as shown in plan (Fig. 84). If frames with 15½-in. top bars are used, as is sometimes done, a simpler body box will be sufficient, as shown in the isometrical sketch (Fig. 86), where it will be seen that the outside pieces B (Figs. 84 and 85) are dispensed with, their places being taken by small pieces fastened on the top edge of the box ends. As the end pieces do not reach the level of the top edges of the sides, a rebate is left into which the ends of the top bar of the frame will fit so as to suspend the frame vertically in the hive.

HIVE FOR REARING QUEEN BEES. 93

The hive body being made to correct size—and as a precautionary measure it is well to have a bar-frame handy to try the fit from time to time during the making, so as to avoid mistakes—the base and roof are made to correspond. Supposing the sides A (Fig. 84) are made of wood planed up to $\frac{3}{8}$ in. thick, the width of the base-board will be 8$\frac{1}{2}$ in., and if the ends B are of $\frac{1}{2}$-in. stuff, the length of the flat part on which the body of the hive rests will be 18 in.; and allowing 5 in. for the flight-boards before the hive door, the total length will be 23 in.

Fig. 87.—Division-board.

Fig. 88.—Distance Rack.

The construction of the base-board and the roof can be best understood by reference to the figures. The dotted lines in Fig. 84 show the roof in plan. In each of the four inside corners of the roof, pieces of wood of square section (D, Fig. 85) are nailed; these rest on the corners of the hive body, and support the roof in position. The door is cut as shown at E (Fig. 85), and a strip of wood is nailed to the lower part of the sides A (Fig. 84) to keep the hive upon the footboard. These strips are shown in Fig. 83, but, in order to avoid confusion, are not shown in Fig. 84.

When the hive is used for queen-rearing, it would be contracted so as just to hold three frames. A division-board or dummy on each side of the frames will effect this. Fig. 87 shows one of the division-boards, which, as will be seen, is a bar-frame with a thin piece of wood nailed to one side of it. This piece of wood is long enough to fit against the sides of the body-box snug, but not tight, and in breadth it is equal to the depth of the frame. In order to lessen the risk of crushing the bees when the frames are moved, it should not reach the floor-board, and the bottom bar should be taken off or a hole bored in it to permit bees chancing to get between the division-board and the hive side to escape.

The frames are kept the proper distance apart

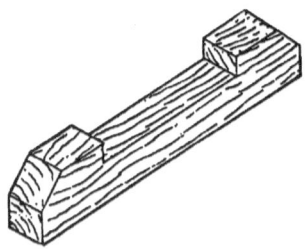

Fig. 89.—Foot of Hive for Rearing Queen Bees.

by means of two racks, one of which is shown in Fig. 88. These are not required if they are fitted with the metal ends bee keepers generally use for this purpose; but it is not difficult to space the frames without the use of either. A few pencil-marks on the hive ends may be used as a guide to accurate distancing.

A stand for the hive may be made by cutting two pieces of wood to the form shown in Fig. 89; the floor-board of the hive rests within the part cut out of the upper surface. Frames for hives are, generally speaking, best got ready-made from dealers in bee keepers' requisites, as they are made accurately to size by machinery, and are sold at a low price.

Hive for Rearing Queen Bees.

In marking out the timber, care should be taken as far as possible to avoid knots coming near the edges of the pieces or in the roof-board, and the heart side of the timber should always be on the outside of the hive. If this is not done, the hive when exposed to the weather is fairly certain to open at the joints. Over the roof-top a piece of calico, cut to size, should be stretched, folded round the edges, and secured by tacks underneath. A coat of thick oil paint upon the calico will make the roof watertight, and the hive itself should also receive a coat of paint. A light colour is best, as dark shades absorb the heat of the sun, making the hive intolerably hot for the bees, and perhaps melting their combs down into a confused mass.

CHAPTER X.

SUPER-CLEARERS.

The operation of removing the honey from the hive and ridding the hive of bees, although, until quite recently, the one operation dreaded by bee keepers, can now be performed with little or no disturbance in as many minutes as formerly the operation required hours, and without the infliction of a single sting upon the operator if ordinary care be taken. This is effected by the use of a super-clearer, an American invention.

The most simple form of clearer is a cone made of perforated metal, fixed in the gable, or gables, of the hive roof, as shown by Fig. 7, p. 17. In use it is simplicity itself. The super to be cleared is gently prised up, small spills of wood (match ends will do) inserted at the corners, and through the orifice thus formed a few puffs of smoke are blown into the hive. After waiting a moment the operator should raise the super, whilst an assistant places a quilt over the body-box or super below, when the full super may be replaced above it; the covering quilt of the super is then removed and the roof put on. The bees, finding communication with the hive proper cut off, make tracks for home through the cone, and once out they cannot return. Stray bees on the prowl (for bees are inveterate robbers) also fail to effect an entrance. By reason of the perforations in the cone, the bees are attracted to its base, where they fail to gain admission; and robbers can find their way inside only when they are sufficiently numerous entirely to cover the cone.

For the reason last given, the cone as a clearer

is unsuited for use in late autumn, or at any other season when honey may be scarce; but in late summer, or when nectar is still plentiful, its use is advantageous. If the super is disconnected from the hive in the early morning it can be left to clear itself during the day; and in the evening, the honey, unaccompanied by a single bee, can be removed indoors.

Some cones are fitted with a delicate steel spring, which, whilst not impeding the egress of confined bees, effectually stops the ingress of any intruder; the cones are also sometimes used

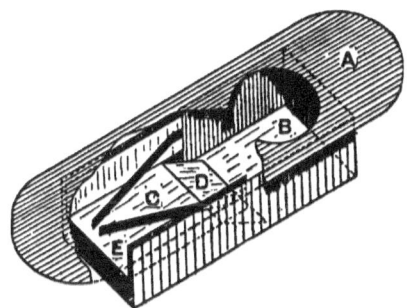

Fig. 90.—Porter Bee Escape.

double—one within the other, ¼ in. apart—for the same purpose.

Another method of using the cones is, instead of entirely uncovering the super of honey, to replace the quilt by a board furnished with several round holes, over each of which a cone is laid or fixed. Escaping bees have thus to pass a double trap, whilst double obstacles are placed in the path of marauding bees. Cones have also been tried the reverse way—that is, to make the bees return to the hive proper without passing into the open air—but have not been much of a success; and practice has proved that the slighter the connection between the hive and the super to be cleared the more quickly will it be rid of bees. The ideal cone should be wide at the base, about 3 in. in length, and have the aperture at the point large enough to pass two bees simultaneously.

However, when by the use of the cone clearer, owing to the lateness of the season or otherwise, robbing is likely to be induced, it is safest not to rely upon it at all, but to use a clearer that affords the bees a direct passage back into the hive without the possibility of return. This is found in the Porter bee escape, obtainable for a shilling of any dealer in bee goods. Fig. 90

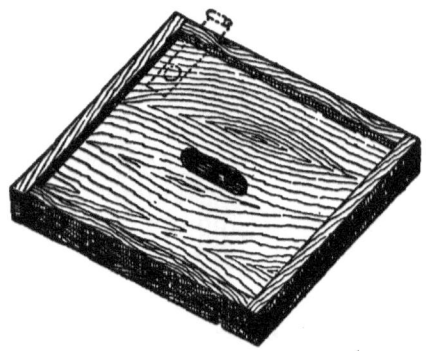

Fig. 91.—Super-clearer Complete.

shows it with a portion cut away to expose the interior. It consists of an oblong piece of thin tinplate A, in size $4\frac{1}{4}$ in. by $1\frac{3}{4}$ in., with a 1-in. hole punched through near one end. To its underside is soldered a rectangular box B, $2\frac{3}{4}$ in. long, $1\frac{1}{8}$ in. wide, and $\frac{1}{2}$ in. deep, one end of which is open. Inside this box a ⌊‾‾⌋-shaped piece C is fixed, 1 in. long, 1 in. wide, and $\frac{1}{4}$ in (full) deep, and the inner end D is bent downwards to meet the bottom of box B. To the inner sides of piece C are soldered two fine springs E of brass ribbon, $\frac{1}{16}$ in. wide, which are bent inwards, as shown, until they almost meet. To an outgoing bee these springs offer little or no resistance; to an ingoer they offer an impenetrable barrier. If the bee, foiled at the apex of the triangle formed by the springs, tries to force a passage by their sides, the only result is that the springs are pressed the closer together, so that to gain an entrance is an impossibility.

With regard to the use of the super-clearer on a hive, it may be said that its size depends on that of the hive, and that no definite measurements can be given beyond saying that appliance dealers usually make it about 16 in. square. Take sufficient dry pine or other wood, ½ in. thick, and joint it to the required width—preferably by grooving, as should it shrink sufficiently to allow a bee-space between the joints its efficiency would be lost—and on each side fix a border, as shown in Figs. 91 and 92, of 1-in. by ⅜-in. wood. In its

Fig. 92.—Section of Super-clearer.

exact centre cut a hole to take the rectangular box B of the escape, into which hole it should fit firmly without the need of further fixing; and in order to allow the bees to escape, the end of the hole at the outlet end must be bevelled off, as shown at H in Fig. 92, which is a section of the clearer.

In use, the operation of removing surplus honey is the same as when using the cone, with the exception that the clearer, instead of a quilt, is interposed between hive and super, the quilts above the super remaining intact. The object of the border round the clearer board will now be apparent: a bee-space is provided above and below. The bees, finding themselves practically cut off from below, will soon discover a passageway out, and, passing through the hole in the top of the escape, they will be guided past the springs and so into the hive below, the stream of bees only ceasing when all have cleared out. When this is the case, the honey may be removed; the honey can be extracted from the frames at once, and the combs returned to be refilled, or cleaned out ready for storing away. In the former case the clearer must be removed; in the latter, by fitting the board with a little extra contrivance, the bees can be re-admitted without disturbance; and, when the combs have been cleaned out dry, the super can be again cleared of bees in readiness for its final removal.

Holes from 1 in. to 1½ in. diameter may be bored through the clearer-board in one or more

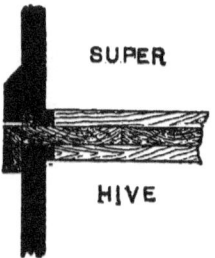

Fig. 93.—Clearer in Use between Hive and Super.

corners, as shown in Fig. 91, and covered with slides of tinplate or sheet zinc, as guides for the working of which tacks or small nails will suffice. In ordinary use the holes are closed by the slides; but when the dripping combs are returned from the extractor, and all is covered up snug, the slides may be withdrawn, allowing the bees free entrance, an invitation they will not be slow in accepting; and if the combs are returned in the evening, the slides may be again closed next morning, when the bees will again pass below through the escape, permitting the now dry combs to be removed later.

In cases where hives are made perfectly square to permit the combs to be hung at right angles to, or parallel with, the entrance, and the frames of successive bodies or supers to be placed at right angles to those immediately below, the clearer board may be the same size as the outside measurement of the hive, and may be provided with a plinth (see Fig. 93) to keep it in place during the period of its use and so prevent undue loss of heat. In Fig. 93, the hive walls are shown in full black lines, the super-clearer being hatched.

The method adopted by a well-known honey producer for removing supers may be given with advantage, as by its adoption smoke—which does not improve the flavour of honey if applied too freely—is unnecessary. Dip a square of unbleached calico into diluted carbolic acid (1 oz. to a pint of water), wring it out as dry as possible, and place the clearer on a stand by the hive. Gently prise up the super as before described, and, shaking out the carbolised cloth, hold it in the hands whilst removing the crate of honey, and by the same movement drop it over the frames beneath, which will cause every bee to disappear rapidly. Place the super on the clearer, quickly remove the cloth, and put clearer and super on the hive, to be left until clear of bees. The same operation can be gone through when taking the clearer off, but on no account must the carbolised cloth remain near honey for any length of time, or the flavour of the honey will be spoiled.

CHAPTER XI.

BEE SMOKERS.

This chapter will describe how to make two kinds of smokers—the Bingham and the Clarke. A Bingham smoker can be made at a cost not exceeding one shilling if the worker can use tinsmiths' tools. These include snips for cutting the thin tinplate; a hatchet stake for turning over the metal for wiring edges or making joints; a large soldering bit; mallets and hammers; punches; and odd pieces of iron.

A hatchet-stake may be improvised from a 2 ft. length of so-called half-round iron, $1\frac{3}{4}$ in. wide, and $\frac{3}{8}$ in. thick. The edges are smoothed with a file, and the iron is supported in a vice.

As to materials for the construction of the smoker, get a piece of best quality tinplate, 12 in. by 18 in., wood, leather, a small bit of $\frac{1}{2}$ in. brass tube, and about 3 ft. of hard brass wire 16 gauge.

Fig. 94 is a general view of the smoker complete. It consists of fire tube, T; funnel, F; handguard, G; strip to secure tube to bellows, H; and bellows, B. The fire tube is $2\frac{1}{2}$ in. in diameter and $6\frac{1}{4}$ in. long. The piece of tinplate should be cut accurately square, $8\frac{3}{8}$ in. by $6\frac{3}{8}$ in. The two short edges are then turned over a little more than $\frac{1}{8}$ in. from the edges, one being turned up and the other down. The piece of tin is bent into a cylinder, and the bent edges hooked into one another and hammered down tight, using a piece of thick round iron or steel as a stake on which to hammer. Run a little solder along the joint to strengthen it.

When the cylinder has been made fairly circular, it will be found to be $2\frac{1}{2}$ in. in diameter.

BEE SMOKERS.

Both ends of this cylinder ought now to be quite flat; but if they are not, the file should be used until they are made so; $\frac{1}{8}$ in. at one end must now be turned out all round at right angles to the body of the cylinder, and this can easily be done with the good tin being used by means of a hammer and the stake, or the sharp edge of a cast-iron lathe bed, which is more solid. The cylinder will thus be reduced to its final length, $6\frac{1}{4}$ in. Within $\frac{3}{4}$ in. of the flanged end, a $\frac{1}{2}$ in. hole is punched through the tin.

This hole is to be coned inwards, as shown in

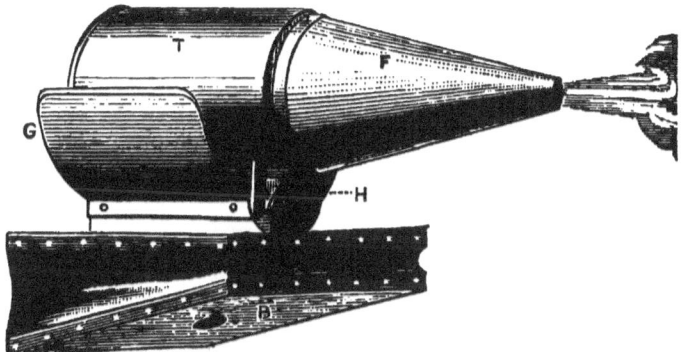

Fig. 94.—Bingham Bee Smoker.

Fig. 95, the object being to direct any of the blast which might impinge upon the sides of the hole inwards into the smoker, rather than between the fire-box and guard. The cone can be shaped with the pane of a light hammer.

The end or bottom E (Fig. 95) is of tinplate, its radius being just $1\frac{1}{2}$ in., $\frac{1}{8}$ in. of the edge being turned up all round, like a cover of a canister; the flanged edge of the cylinder is laid in it, and the edges turned in to embrace the flange and keep all tight, as shown at E (Fig. 95). The edge of the bottom can be turned up on the end of a piece of thick round iron.

The funnel may next be made, its pattern being

104 *Beehives and Bee Keepers' Appliances.*

shown by Fig. 96. The arcs of circles should be scribed on the sheet of tinplate, the inner being 1⅜ in., and the outer 6⅞ in. in radius; measure off

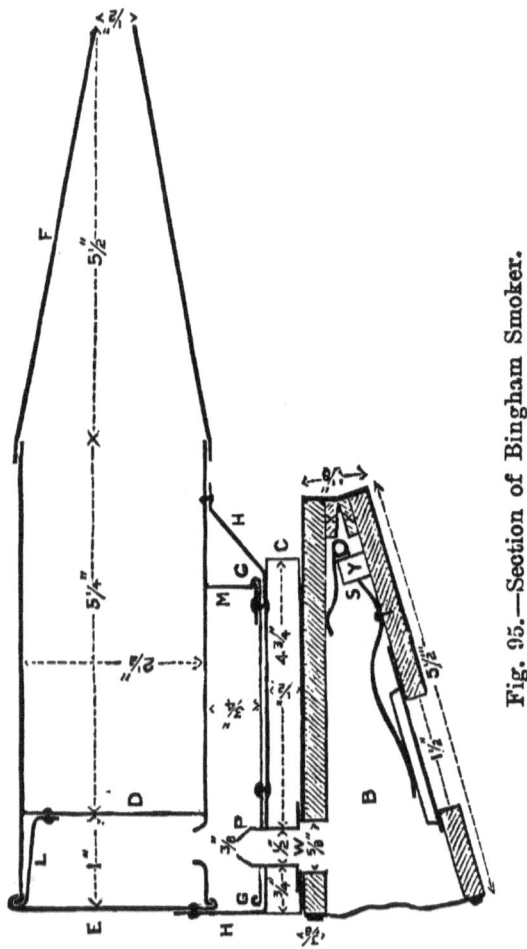

Fig. 95.—Section of Bingham Smoker.

8⅜ in. on the circumference of the outer arc, and draw lines to the centre; the piece may then be cut out, turned up and down at the edges, and connected in the same way as the body tube. Do not, however, turn down quite so much at the edges so as to make the large end of the funnel big enough to embrace the tube T (Fig. 94 and 95). The end of the funnel should be hammered, so as to make about ½ in. of its wide end parallel to fit on the cylindrical body. The body of T could be tapered very slightly to assist the putting on of the cone.

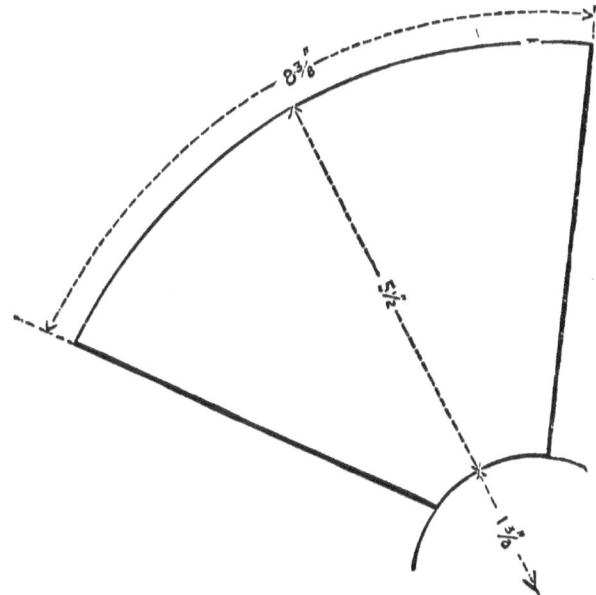

Fig. 96.—Pattern of Funnel.

The piece M is a sort of additional support to the barrel. It is simply a piece of tinplate 1½ in. square, having two edges turned down and hammered flat, making it ⅞ in. wide. It is then bent at right angles, having one leg ⅝ in. long, and the other ⅞ in. long. It is attached to the hand-guard and support H by the short leg, while the long one is hollowed out to fit the curve of the body of the blower.

The hand-guard is a piece of tinplate, No. 16

B.W.S., 6⅞ in. by 4⅜ in. wired all round with thin wire. The wiring is easily done by first cutting small pieces off the corners of the tin, then turning up the edges all round, laying the wire in the edges, and then hammering down so as completely to cover the wire, and leave a nicely formed bead. The hand-guard is bent to a semicircle.

The support is made of a strip of tinplate 7½ in. long and 1⅝ in. wide. Lines must be scribed on one surface ⅜ in. from each edge, thus marking it into three parts, the centre one being ⅞ in. wide. At 1¾ in. from one end, and at 1¼ in. from the other, nick the sides into the lines, and turn over

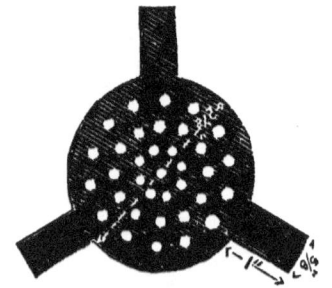

Fig. 97.—Smoker Diaphragm.

the edges of the end pieces, thus doubling the tin at the ends and making them only ⅞ in. wide. The centre must also have its edges turned up, but only at right angles, to the middle part, thus making a sort of trough which will fit on the piece of wood s (Fig. 95) nailed to the bellows board. The ends of this piece of tinplate are now turned up in the opposite direction to the trough, the short one at right angles, and the longer one at an angle of 135° with the middle part. This last end is to be bent again, a little more than 1 in. from the first bend, so as to lie parallel to the centre part.

For riveting these parts together, obtain four rivets ⅛ in. thick and ¼ in. long. One rivet connects the support H, the hand-guard G, and the little piece M, all together towards the front, while another 1½ in. from the back holds H and G together. A ⅝ in. hole should now be made in H and G ¾ in. from the back, and the whole may be attached to the barrel with the two rivets shown. The rivet holes are punched. The top of M should be filed hollow to fit the curve of the barrel, the front part of H being hammered to a similar curve.

The diaphragm D (Fig. 95) is of sheet iron, and the legs are sometimes riveted on, but it is easier to cut it out of the iron in one piece as in Fig. 97. It should be punched with ⅛ in. holes. The diameter is just a little less than 2½ in., and the legs are 1 in. long; they are turned up at right

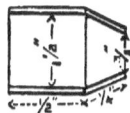

Fig. 98.—Coned Blast Pipe.

Fig. 99.—Nicked Tube for Making Blast Pipe.

angles to the body. The leg L (Fig. 95) is riveted on.

The only difficulty likely to be experienced in the coned blast pipe P (see also Fig 98) is the coning of the mouth to ⅜ in. This, however, is easy after cutting two nicks in the piece of ½ in. tubing as in Fig. 99. Anneal the brass by heating it in the fire, and when cool, the point can be hammered cone-shape without much trouble. A touch of solder will mend the cut afterwards. This is not a professional method, but is quite good enough for the present purpose.

The piece of wood S (Fig. 95) bears the whole weight of the tin portion. Oak will be found the best material; and suitable dimensions are 4¾ in. by ⅞ in. by ½ in. At ¾ in. from one end make a ½ in. hole to take the blast-pipe; the upper edges are to be bevelled off and hollows cut to take the heads of the rivets in H (Fig. 95).

The boards of the bellows can be made of any tough and thoroughly seasoned wood. Two pieces for the cheeks of the bellows are 5 in. by 5½ in., two other pieces X (Fig. 95) are 4⅞ in. by ½ in., ⅛ in. thick at the back and rather more at the front, so that when the backs of the two boards are brought together the front joint will not open. The piece Y is ⅜ in. square and 2½ in. long, and another piece is $\frac{3}{16}$ in. square and 3½ in. long. The wood piece for the valve is 2 in. square and ¼ in. or so thick. When all the pieces have been nicely planed and rubbed on a sheet of glasspaper, bore a hole for the valve in one bellows board and one at W for the blast pipe in the other. The latter is ⅝ in. in diameter, and ¾ in. from the back end of the board; the former is 1½ in. in diameter, and its centre is 2 in. from the back end of the board. The burrs which may have been formed in the boring of these holes should be carefully glasspapered off. The pieces X should then be glued and tacked across the front edges of the boards, and the piece Y similarly fixed on the lower one, ⅜ in. behind X, its ends being equidistant from the edges of the board.

The springs (Fig. 100) must next be made; each is a piece of wire bent into two parts, and as there are two springs there will be, in effect, four wires pushing the boards apart. To make the springs, drive a couple of wire nails $\frac{3}{16}$ in. thick into a piece of wood 6 in. apart. Cut the 16 gauge wire 14½ in. long and straighten it, place it against the wire nails with the ends projecting equally at both sides, and turn the ends round the nails, one to the right and the other to the left. Give two and a quarter turns to each end, which will leave them at right angles to the middle part as in Fig. 101. Then bend the middle part at M into a curve so as to bring the coils together and the loose ends

BEE SMOKERS. 109

lying side by side. Then with pliers turn down the points to prevent them from sticking in the boards, and give a little bend just near the coil (see Fig. 100). The ends of the piece of wood which have been prepared, $\frac{3}{16}$ in. square and $3\frac{1}{2}$ in. long, are rounded and passed through the coils of the springs, and a little bit of thin wire ties them together to prevent their slipping off. This axle, as it may be called, of the springs is then placed between the pieces x and y (Fig. 95), and the lower board then resembles Fig. 102, this illustrat-

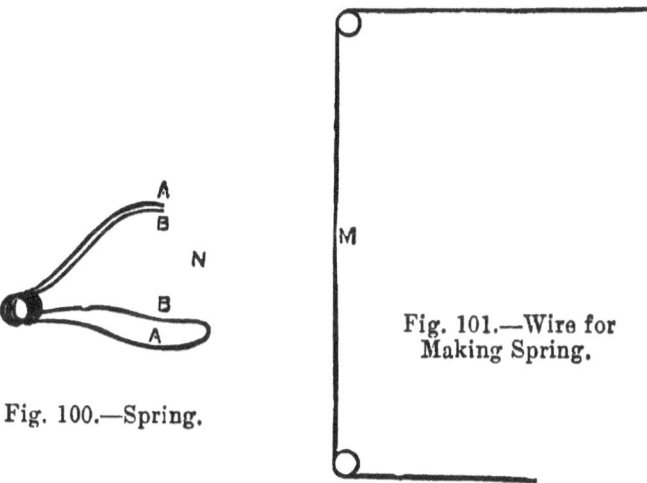

Fig. 100.—Spring.

Fig. 101.—Wire for Making Spring.

ing the bottom board of the bellows with springs in place.

The valve is a piece of leather 3 in. by 2 in., to which the 2 in. square piece of wood is secured at one end by a tack passing through near its centre. If it were glued to the wood, and the latter warped, the valve would not close. Three edges of the wood valve will coincide with three of the leather, and an inch of the leather will project beyond the wood. By means of this tongue the valve is secured to the board with two tacks. A light spring (shown in Fig. 95) presses very

110 *Beehives and Bee Keepers' Appliances.*

gently on the back of the valve to prevent it from opening except under suction. This spring can best be made of a bit of watch-spring, but a bit of thin hard brass wire does almost as well; a tack passing through a hole in the watch-spring, or a loop turned on the end of the wire, will fasten it to the bellows board. It would be well to put a narrow strip of leather over the valve and fasten its ends down with two tacks, allowing the valve only about ⅛ in. rise. This is to prevent mis-

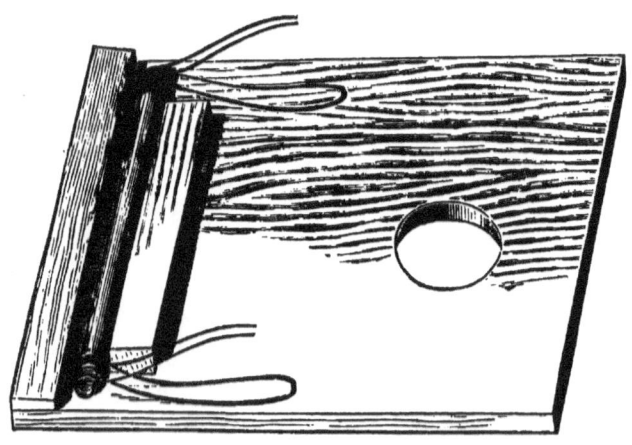

Fig. 102.—Bottom Board of Bellows.

chievous persons thrusting odd articles into the bellows.

The hinge may now be tacked along the front edges of the boards and of the pieces x (Fig. 95). It is a strip of leather 5 in. by 1½ in., and should be glued as well as tacked. The springs may be put in place, and the outer edges of the bellows boards held 2¾ in. apart while the leather is being glued on. The edges of the boards are rubbed over with strong glue, and the leather laid on and secured with short tacks 1 in. apart. The edges may afterwards be cut flush with the outside of the bellows boards.

As a general guide in procuring the leather it may be said that a strip 18 in. long, tapering from 3¼ in. in the middle to 1⅛ in. at the ends, does for the three sides, an inch being left at each end to overlap the hinge. When the leather has been put on, the bellows ought to be able to work nicely; but, for the sake of appearance, it is customary to put a narrow strip of red leather

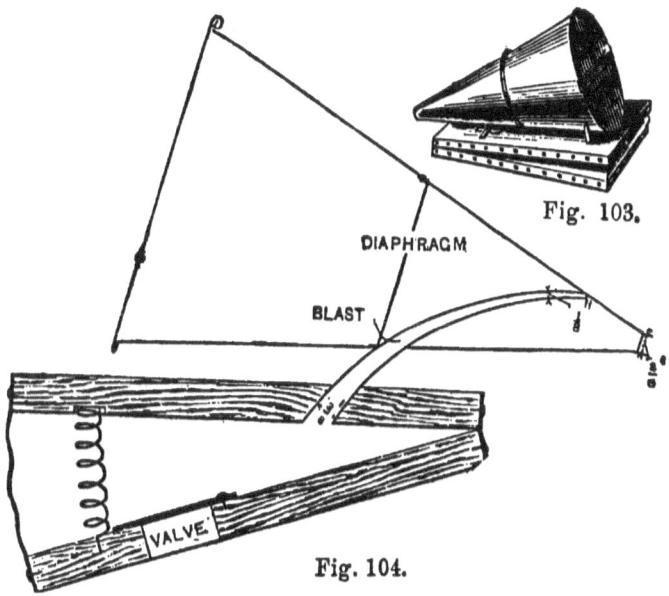

Fig. 103.—Clarke Smoker. Fig. 104.—Section of Clarke Smoker.

all round the edges of the boards, and to secure it with small brass tacks.

Before the bellows are put together it will be much better to fasten to the upper board the piece s with glue and a couple of screws from the inside, the blast holes in both bellows and support being over one another. A piece of wire-gauze w (Fig. 95) should be put between the two, covering the blast hole in such a way as to prevent ash or cinder from the smoker finding its way into the bellows. The contracted piece of brass pipe can now be pushed into place, and the body of the smoker fastened to s with the piece of tinplate H, which has its edges turned down so as to embrace it; four little screws or tacks will hold it on very firmly.

The smoker is now finished and ready for the fuel, which can be brown paper, sacking, or anything that will smoulder. It should be inserted in such a way as to afford passages for the air through it. If it is packed tightly the smoke cannot be expected to travel through the entire length of the barrel.

The form of spring used in the real Bingham smoker entails a good deal of unnecessary labour, and probably a spiral spring of the proper strength could be substituted. Then the front edges of the bellows boards could be brought together, and the four pieces of wood there found could be dispensed with.

The Clarke smoker is simpler than the Bingham, but hardly so efficient. Fig. 103 shows it in general view, and Fig. 104 is a sectional view. The bellows boards are $4\frac{1}{2}$ in. by $6\frac{3}{4}$ in., and nearly $\frac{1}{2}$ in. thick. A 1-in. hole is made through one board, its centre being $2\frac{1}{2}$ in. from one end of the board, and midway across it. This is for the valve, which is simply a piece of stout leather nailed on one side and free to rise on the other, after the manner of a butterfly valve. At a distance of $1\frac{3}{4}$ in. from the opposite end of the other board, a $\frac{1}{2}$-in. hole is bored, sloping from the front as shown in Fig. 104. This is to take the blast pipe. The boards are kept apart by a strong steel spiral spring, which is placed at the side of the valve, and more towards the back, just where the pressure of the hand goes. The boards at their widest part are 3 in. apart, and at the narrowest, 1 in.

BEE SMOKERS. 113

The entire bellows of the Clarke smoker could easily be made by cutting out and planing the boards, boring the holes, and tacking on the valve; then the points could be brought together, and a slip of leather, 4 in. long and 1 in. wide, tacked along them. The spiral spring, which could be made of hard brass wire, No. 18, B.W.G., could then be put in place, being fixed to the boards, either with a straight piece of wire left at both ends of the spring, or the ends of the spring could

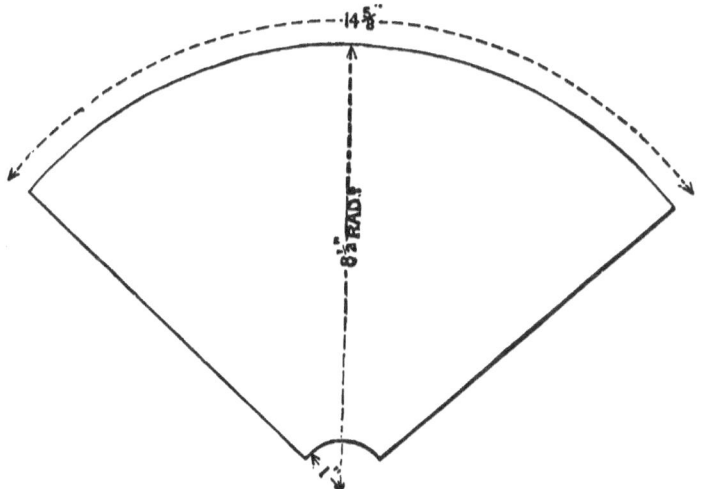

Fig. 105.—Pattern of Funnel.

fit into holes bored partly through the boards. A bit of wire could then be bent so as to keep the boards 3 in. apart at the wide end while the leather was being glued and tacked on. The leather should overlap the piece already tacked to the front by about an inch. Basil leather will answer if better is difficult to obtain.

The fire box of the Clarke smoker should be made of stout tinplate. Its pattern is shown by Fig. 105. The circles there shown should be carefully scribed with a compass on the sheet of tin-

H

plate, and then cut out with a pair of snips. The straight edges should be turned over for ¼ in., one up and the other down, and the piece bent into the shape of a funnel. The parts turned over will then catch into one another, and should have a little solder run along them after they have been hammered tightly together. The wide end of the funnel, for the distance of ⅛ in., is now to be turned straight out so as to take the bottom (Fig. 106), which is attached to it just as a tinman fines a bottom to a can, except that after it is turned over once it is left standing out from the funnel, as seen in Fig. 104.

Before the bottom is fastened, it would be as

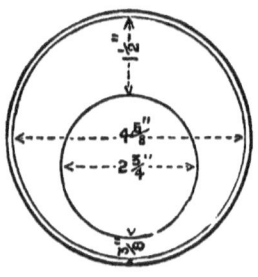

Fig. 106.—Bottom of Funnel.

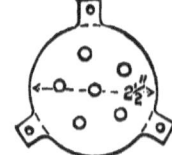

Fig. 107.—Diaphragm of Clarke Smoker.

well to make the tinplate diaphragm (Fig. 107) and fix it to the funnel. It is 2½ in. in diameter, and is punched with a number of ⅛ in. holes. In bought smokers it is fitted in a kind of bead moulded on the funnel; but in the present case three or four projecting tongues are left, and these are turned over and fastened to the funnel with small rivets.

The bottom with its fire door cut out is shown by Fig. 106. The outer dotted line in this figure shows the part which will be turned over to embrace the funnel end. The door is a piece of tinplate large enough to cover the hole and to pivot on the rivet shown in Fig. 103. Its edges are turned over so as not to cut or scratch the hand of the operator.

Bee Smokers.

The tinplate blast pipe is 4½ in. long, its bore tapering from ⅜ in. to slightly more than ⅛ in. The pipe is bent into the shape shown by Fig. 104, and extends to within ⅝ in. of the front of the fire holder. The whole of the sheet metal work is attached to the bellows with two screws, holes for which, 2¾ in. apart, must be punched within ⅝ in. of the base of the funnel. The blast pipe is hooked into the hole in the bellows made for its reception, and the screws are put in at the back, a couple of bits of tinplate tubing, 1 in. long, through which the screws pass, preventing the bellows and fire box coming in contact with each other.

CHAPTER XII.

HONEY EXTRACTORS.

THE first of the extractors to be described in this chapter is of the Little Wonder pattern; it is useful to a bee keeper with two or three hives, but is unsuitable for a larger apiary, as only one comb at a time can be operated on. It is difficult to use, and if the honey is thick from natural causes or through cold weather the extractor cannot be made to revolve at a sufficiently high speed to clear the combs. Where a number of hives are worked for extracted honey, a geared cylinder extractor (described later in this chapter) is necessary.

To make the extractor (of which Fig. 108 is a side view showing the handle raised) a piece of clean, straight-grained red deal, 3 ft. 10 in. long by 1¾ in. square, is required. From one end cut off 7 in., and, beginning at about 6 in. from both ends, chamfer down to circles 1½ in. in diameter, and fix ferrules about 1 in. long as shown at A (Fig. 108). Then insert two pieces of ⅜-in. round iron, one 10 in. long and the other 4 in. long, for about 3 in. into the ends; the long piece should be at the top, and the short piece at the bottom. Take the 7-in. piece of wood, centre the ends, and bore a ⅜-in. hole through from end to end and round the wood to 1½ in. in diameter as shown at B (Fig. 108). Next bend two pieces of 1-in. by ⅛-in. hoop iron about 2 ft. 9 in. long to the shape shown in Fig. 109. It will be well to make a rough template of the shape, and, if the iron is fairly good, the bending can be done with a hand vice after heating the metal in the fire. Place

HONEY EXTRACTORS.

these irons at C (Fig. 108), the method of fastening them being shown in Fig. 109.

For the can a piece of tinplate, 1 ft. 9½ in. long by 1 ft. 3½ in. wide, will be required to form the back; the template used for bending the irons may be reduced ¼ in. and used in shaping the tin as shown by Fig. 110. Four cleats A (Figs. 110 and 111) about 6 in. long, and two smaller ones B (Figs. 111 and 112), of stout tin, are soldered to the sides to hold the cage in position. Two pieces of tinplate, cut to the shape shown by Fig. 113, and allowing a ⅜-in. margin round the outer edge for the joints, are required for the top and bottom; the semicircular hole A (Fig. 113) is required in the top piece only. These pieces can be fixed to the back, and the joints folded and soldered. A piece of tinplate 9½ in. by 3¾ in. is required for the front C (Fig. 112); the joints are turned over and soldered at the bottom and sides.

The front and bottom may be worked in one piece by making the bottom 7 in. wide instead of 4 in. as in Fig. 113. The opening in the front should be strengthened with No. 11 B.W.G. wire

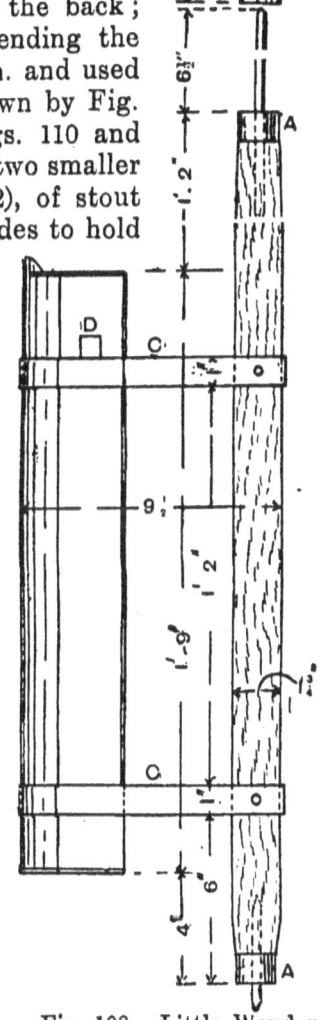

Fig. 108.—Little Wonder Honey Extractor.

run round, the tinplate being turned over it as at c (Figs. 110 and 111).

To complete the can, cleats to hold it in posi-

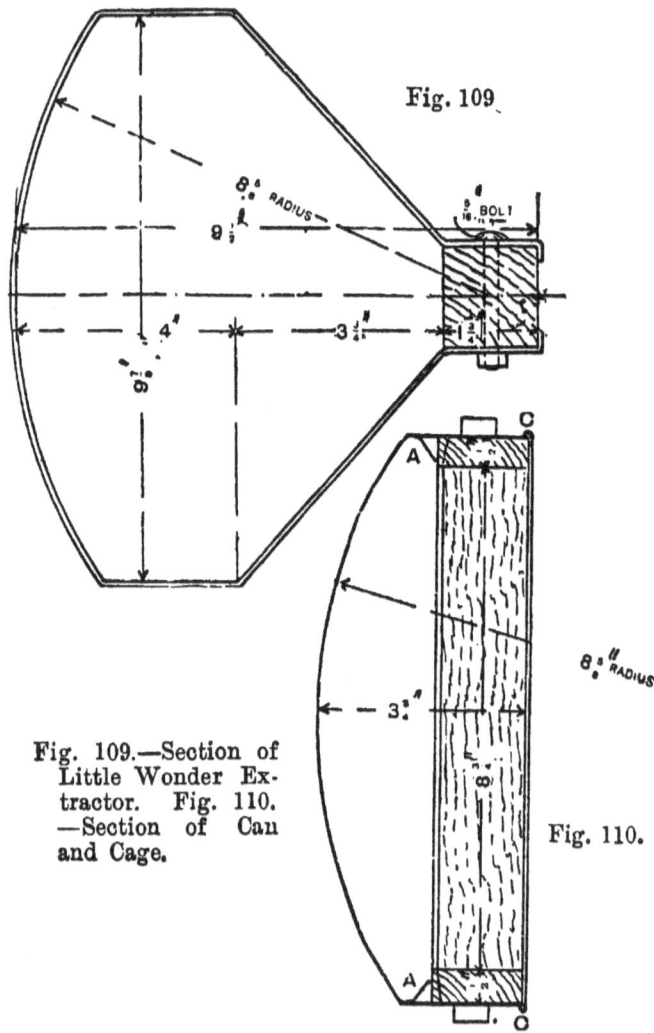

Fig. 109.—Section of Little Wonder Extractor. Fig. 110.—Section of Can and Cage.

tion when dropped into the frame are fixed at D (Figs. 108 and 112), and a lip is soldered to the top for pouring out the honey. Good stout tinplate

HONEY EXTRACTORS.

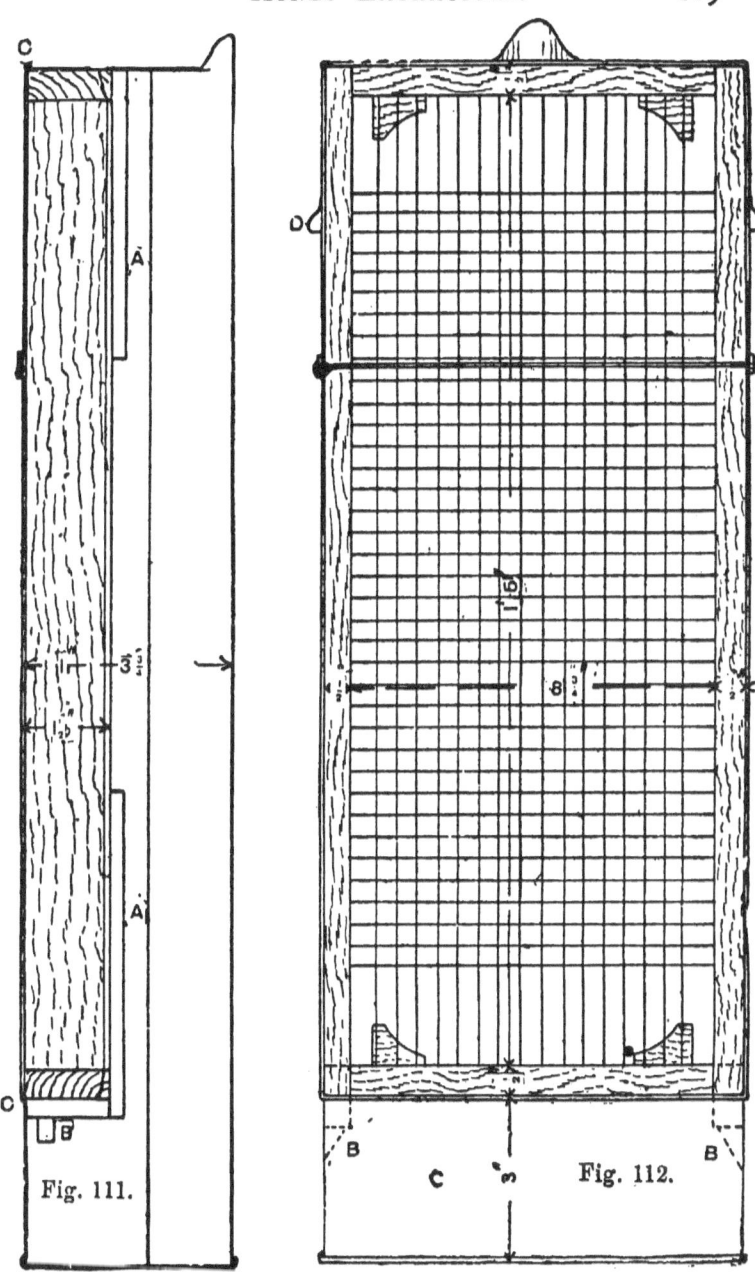

Figs. 111 and 112.—Section and Front View of Little Wonder Can and Cage.

120 *BEEHIVES AND BEE KEEPERS' APPLIANCES.*

should be used in making the can, or it will collapse when a heavy comb is being extracted.

To hold the cage in position when the combs are in, a wire is placed across the front at E (Fig. 112); one end is fastened with a small wire staple soldered to the cage, and the other with a catch of double tin or wire.

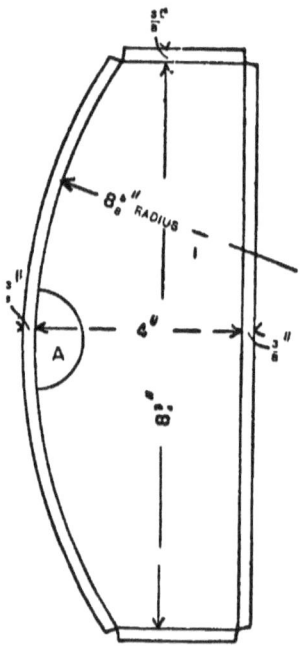

Fig. 113.—Pattern for Top and Bottom of Extractor.

The cage is of 1½-in. by ½-in oak or beech of the dimensions given in Figs. 108 to 112. The joints at the top may be dovetailed, and at the bottom, mortise-and-tenon joints may be used. A gauge line is run round the inside of the frame ⅛ in. from the bottom edge, and holes about ⅜ in apart are bored with a fine bradawl. The holes in the sides should commence about 1¾ in. from the ends, and a space of ⅝ in. should be left at each side

HONEY EXTRACTORS.

to allow the metal ends of the frames to pass through and bring the combs close to the wire of the cage. The holes should be bored at an angle as shown at A (Fig. 114) so as to strengthen the cage. No. 20 B.W.G. tinned wire should be threaded through the holes as shown in Figs. 112 and 114, the long wires being put in first and the cross wires interlaced with them. Galvanised netting is sometimes used for the cage, but this is objectionable, as zinc should never be allowed to come in contact with honey. Suitable wire can be obtained from makers of bee keeping appliances.

Wooden cleats are glued and nailed at each of

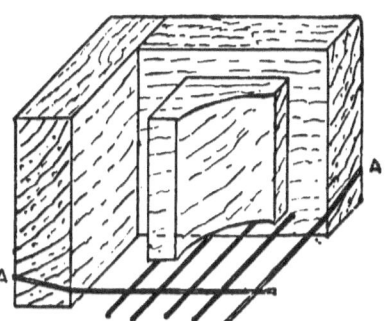

Fig. 114.—Corner of Cage of Little Wonder Extractor

the corners to keep the ends of the frames in position when extracting; the forms and positions of these are shown in Figs. 112 and 114.

An iron plate 3 in. square by $\frac{3}{8}$ in., with a $\frac{1}{2}$-in. sinking in the centre, should be screwed at the corners to the floor of the room where the extractor is being used. The spike at the bottom of the extractor will work freely in the plate, and will prevent the spike slipping.

The same principle governs the action of the Little Wonder and cylinder extractors, the honey leaving the cells of the comb by centrifugal force, but, while in the Little Wonder the entire machine revolves, carrying comb, receptacle for honey, and any honey which has already been extracted, in the cylinder extractor as few parts as possible are made to revolve, and this is a decided advantage. The labour necessary to overcome the inertia of a large mass of material in starting and stopping is saved, and the decreasing weight of the comb, which is perceptible when it and its connections only are revolved, is an indication to the operator that the honey has been extracted.

For a given velocity, the nearer the comb is to the centre of revolution the greater will be the

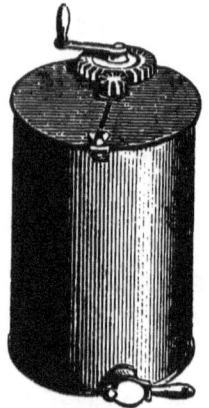

Fig. 115.—Cylinder Honey Extractor.

Fig. 116.—Cylinder Extractor with Frame Inside.

centrifugal force, but the honey in most of the cells will tend to press against their sides as well as leave them. To overcome this tendency the combs should be placed at an infinite distance from the centre of revolution. It is evident that practice more than theory is what will decide the best position for the combs, and from exhaustive experiments, Mr. Cowan has concluded that the outer surface of the comb should, during extraction, be placed 6 in. from the centre of the spindle round which it revolves. He has also decided that extractors which hold two combs at the time are preferable to those which hold four or more.

HONEY EXTRACTORS.

The cylinder extractor, then, consists of four distinct features: (1) the frame which holds and carries the comb baskets, (2) the comb baskets, (3) the cylinder, or barrel in which they revolve, and (4) the driving gear, or crank. Fig. 115 is a general view of the extractor; Fig. 116 shows the extractor complete with the frame inside, the baskets being in place; and Fig. 117 shows the frames, baskets, and crank handle.

For the frame forming the first item in the list, three sheets of tinplate, 17 in. by 12½ in., are required. One of the sheets is cut into six strips 2 in. wide. The edges of tinplates are not always true when they come from the shop, and should

Fig. 118.—Wired Tinplate.

Fig. 117.—Frames, Baskets, etc.

therefore be pared until straight. Three strips should be joined together end to end, by turning ¼ in. at the ends over, hooking together, hammering down flat, and touching with solder, as before described. These strips may now be cut to 45 in. in length each, and wired at both edges with wire about ⅛ in. in diameter, about No. 10 gauge. The wires are to be 44 in. long each.

The wiring is effected by turning the edges of the tinplate over for a distance of nearly ½ in. by means of a mallet and the stake; the wire is then laid along the trough thus formed, and the edge of the metal hammered down so as entirely to envelop it. A good deal of tapping and some practice are required to make a neat bead.

As the cage is likely to be often smeared with

honey, which gets in between the wire and tin, and there sets up fermentation, or becomes a constant source of dirt, the following plan for preventing this may be adopted if desired: Solder the tin along outside the wire, so that a nicely formed hollow is made, which can easily be kept clean, and has no corners for dirt (see Fig. 118). In this, as in every other part, use fine solder containing a large proportion of tin.

It is best to do all wiring while the tin is in the flat, not after it has been bent into shape.

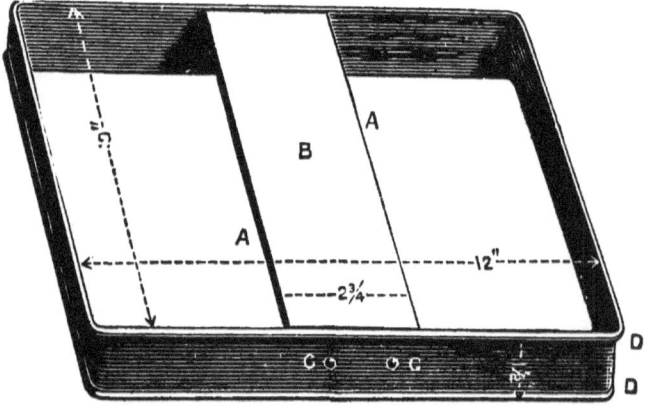

Fig. 119.—Rectangular Band and Bridge.

The wire can be soldered in either before or after the bending, but it is easier to do it before. The tin is longer by 1 in. than the wire, but this should not be turned over until a later stage of the work is reached.

These wired strips of tin are now to be bent so as to form two rectangular bands, 12 in. by 10 in.; the overlapping inch at the ends forms a good strong joint when thoroughly soldered. The bending can be done with a wooden vice, such as is usually found on a carpenter's bench, and should be as nearly as possible to a right angle, and the frame or band should not be in winding, but lie

HONEY EXTRACTORS. 125

flat on the bench; the wired edge should be turned outwards, leaving the inside surfaces flat.

Another sheet of tinplate must now be taken in hand, and two pieces, 12 in. by 3½ in., cut off across it. These must be wired at both edges, with wires only 10 in. long placed in the middle, leaving 1 in. at each end free of wire; but the turned over edges of tin may be hammered down flat at the ends. The unwired ends may now be

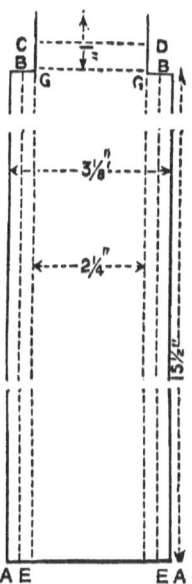

Fig. 120.—Pattern of Slide.

turned up sharp where the wire terminates, thus forming a sort of tin stool 10 in. long, and with legs 1 in. high. These are to be fastened with solder and small rivets to the bands already made. They will, of course, bridge the band the narrow way, which is the only direction in which they will fit, and be equidistant from each short side. Fig. 119 is a diagram of one of the bands and bridges at its present stage. The bridges are for the purpose of attaching the spindle.

For the slides, cut four other pieces of tinplate, 16½ in. by 3⅛ in., and turn over and hammer down ¼ in. slack at each long edge of the entire lot; turn up also ½ in. at one short edge, or end, and hammer down flat; but before any of this turning down is done it would be advisable to cut out rectangular pieces at the corners, so as to prevent the tin from being doubled too much. Fig 120 gives a pattern of a slide. The long edges are doubled down along the lines A B, the short one at C D. The long edges are to be then turned up along the line E G, and left standing at right angles to the broad part, and the short end along the line G G, the whole thus forming a kind of trough, open at one end, and having the other end double the height of the sides. A touch of solder in the corners will bind the edges firmly together, and make the work stronger. The sharp corners of the projecting end should also be nipped off, and rounded nicely with a file.

The bands may now be connected together with these slides, into which the comb baskets slip. Place one of the bands on the bench with the bridge up, and stand a slide, with the stopped end down, at one of the corners, its back surface being in contact with one of the long sides of the band, and pushed up as tightly as possible towards the corner. A small cramp and two pieces of wood can be used to hold the two firmly together, while the square is applied to see that the slide is at right angles to the band. A little solder is then run between the two, and a similar operation performed with the three other slides. The upper band can now be put over the ends of the four vertical slides, and if the work has been done carefully, it will be found to fit well, each slide going right into its corner. If, however, things are not true, the square can be applied, and the erring slide or slides found, unsoldered from the lower band, and set right. See that both bridges are turned upwards so as not to form troughs to hold the honey, which they might do if turned the other way. Fig. 121 is a diagram of the work at this stage.

Fig. 121.—Bands, Slides and Bridges of Extractor.

Small rivets could be used as well as solder to hold slides and bands together, but they are scarcely necessary; but if used, their heads must not project into the slides, and hinder the baskets from moving freely up and down.

Instead of this framework, a box 10 in. square and 15 in. high could be employed, having the slides soldered on to two of the sides, while the other two act as backs for the comb baskets. This would be found by many an easier piece of work, and possesses the additional advantage of being easily cleaned, and affording few corners for dirt to lodge in.

With regard to the comb baskets, half of one of the baskets is shown by Fig. 122, which figure

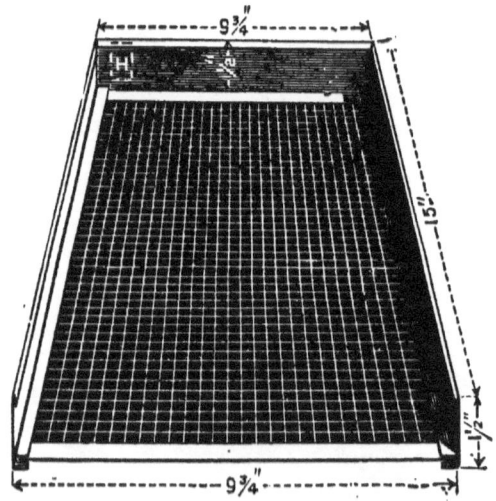

Fig. 122.—Half of Comb Basket.

is drawn from a point near the basket and between the sides. Each basket consists of a bottom of wire netting, two sides of tinplate 1½ in. high, and one end; the other end is wanting, as it will form the top of the basket when in position. The other half basket is exactly similar to this, but a little narrower, so as to fit inside it, as can be seen in the lower part of Fig. 123. The width of the outer half of the basket is such as to fit easily between the slides; it may be 9¾ in. The distance between the two nettings can be varied from 1½ in.

to more than 2 in., as will be understood from Fig. 123.

To make these baskets, the four pieces of netting should first be procured, cut accurately, two to 15 in. by 9⅜ in., and the remaining two ¼ in. narrower. They should then be bound round with tinplate which overlaps ⅜ in. at each side. To do this, lay the straight strips of metal, which will be ¾ in. wide, on the bench, and the edges of the netting over them and halfway across. Then solder each wire to the tin, turn the tin over, and solder each wire to the turned-over part also, using plenty of solder and heat, so as to have every wire very firmly held in. By this means the netting has a metal frame, which will greatly strengthen it, and prevent it from sagging. Strips

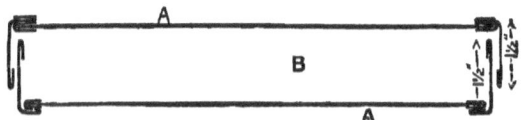

Fig. 123.—Section through Comb Basket.

of tinplate must now be soldered together to form four long pieces 41 in. long by 2¼ in. wide. The edges of these are to be turned over and hammered down to the extent of the usual ¼ in., and one edge turned up at right angles, so that a section of the strips will form an L, one leg of which is 1½ in., and the other ¼ in. Each strip is then to be bent into such a shape as to form the three sides of the half-basket shown in Fig. 122. In two of them the short side is 9¾ in. long, and in the other two ⅛ in. less. To bend the strips it will be necessary to cut the narrow rib with a chisel.

The framed netting can now be laid in position, and soldered firmly against the narrow rib, so that there are four thicknesses of tin round the netting.

About $\frac{3}{8}$ in. of the sides will project beyond the limits of the netting; this, in the wider pair, should be turned over, and a short bit of wire put in it to afford a hold when drawing out the baskets. In the narrower pair some may be clipped off, and about $\frac{1}{4}$ in. turned down, so as to have a nice round edge at the top.

To use these baskets, the comb is uncapped at both sides and laid on one half of the basket; the other half is then placed over the first, which it fits, like the lid of a pasteboard box, and the entire basket and comb is slipped down the slides of the extractor, another comb being put into the other basket and slides. The whole is then whirled rapidly until the honey from one side of the comb is extracted; the baskets are then withdrawn, and the other sides of the combs turned outwards and extracted in like manner.

It is unfortunate that a hole must be cut out of the end of each half of the comb basket, so as to let the long top bar of the frames pass through. This could be avoided, however, by making slides and baskets an inch or so longer.

The spindle is not made until its exact length is known (not until the case is made) yet it is convenient here to describe its construction. Any one of three kinds of spindle may be used; the most workmanlike would probably be $\frac{1}{2}$-in. round iron or steel, tinned all over, or covered with tinplate soldered on, or it might be a tinplate tube, though this is not recommended. In any case, it passes through the bridges at their middle points, or nearly so, in such a position as to make the cages revolve truly and evenly. The lower end is brought to a long cone, and works in metal bearings soldered to the centre of the bottom of the can. The top of the spindle takes either a cranked handle or a toothed pinion, with which it is driven. The tops of the cages should be 2 in. lower than the top of the can.

It will be necessary to put tin washers in the bridges to strengthen the hold of the spindle. They could be 1½ in. in diameter, beaten saucer-shaped, with a hole in the middle, through which the spindle passes. After it has been soldered to the bridges these washers could be placed over the point, and attached both to the spindle and bridges. It would be well to have the holes a little small, and to turn out the edges until the spindle can pass through. This will give a firmer hold to the solder than the mere thickness of the tinplate could afford.

For the cylinder itself, get a tinplate 55 in. by 26 in.; the top and bottom should be wired with ¼ in. wire, and the edges turned over to form a joint. The sheet is then to be bent into a cylinder, and the joint made and soldered. A piece is next to be cut for the bottom, and the edge turned up ¼ in. all round. The bottom, however, should be slightly, say ½ in., larger than the diameter of the cylinder, as it is to be placed in it in a sloping position, so as to allow all the honey to drain out of the cylinder through a treacle valve, which is placed in the lowest position (see Fig. 115). This valve can be obtained from dealers in hive furniture. The flange of the bottom will be turned down, and firmly soldered to the sides of the barrel. The centre of the bottom being ascertained, a bearing for the lower end of the spindle can be soldered in place either before or after the bottom has been fixed; the under surface of the bearing is to be filed to an angle to suit the bottom, so that its top surface is horizontal.

A couple of bands of hoop iron, ⅛ in. thick and 1¼ in. wide, riveted to the edges under the bead, greatly strengthen the cylinder. To the top one, attach the bolts (Fig. 124), which hold the bar forming the top bearing for the spindle, one of the bolts breaking the joint of the hoop. The bar is fastened with fly nuts or hexagonal nuts. This bar is 1¼ in. by ⅜ in.—long enough to reach across—with holes drilled for bolts, and one for spindle to pass through. A plain short crank handle does for driving, it being the simplest and cheapest.

The two wires shown crossing each other in Fig. 1,7 (p. 123) should receive attention. They are to prevent the network from bulging, and are ⅛ in. thick, fastened with solder to the framework at their ends, and to each other in the middle. The outer wire should be bent at the point of juncture, so as to be flat against the network; otherwise, it would be of very little use.

If it is decided to use gearing wheels as shown in Fig. 115, they can be bought cheaply. The pinion

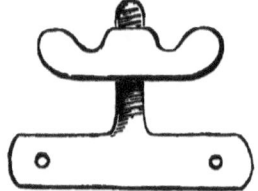

Fig. 124.—Bolt for Cross-bar.

fits the spindle, and is keyed to it, and the toothed wheel works on a stud riveted to the cross-bar. The cross-bar would, in this case, require to be somewhat stronger—say, ½ in. thick.

The extractor is finished by the addition of a couple of handles riveted to the sides, and covers, of which there are two, one at each side of the cross-bar. It is far easier to have the covers flat, in which case the edges can be turned down, and made to embrace the rim which fits into the barrel. Inspection of an ordinary saucepan cover will show how this can be done.

CHAPTER XIII.

WAX EXTRACTORS.

Wax extractors, which follow honey extractors in natural sequence, are not so indispensable as those appliances.

Much wax extracting can be done with a simple milk strainer and a saucepan. The strainer should be about 8 in. in diameter and have a wire netting bottom and sloping sides. The lower part of the strainer should fit into the saucepan, the upper part being supported clear of it. Put water in the saucepan, affix the strainer, put the combs in the latter, and put a cover (that of the saucepan if it fits) over the top of the strainer. The whole is then put on the range, where the water is brought to the boil; the steam will rise through the strainer, and melt the wax, which passes through to the water underneath, leaving any dirt or refuse in the strainer. When all the wax is extracted, the water is poured into a basin, and the wax, when cool, will be found in a cake on top.

This is very simple and inexpensive, and is on the same principle as the Gerster extractor, except that the wax does not there come into contact with boiling water.

The solar extractor produces the best quality of wax. It can be used only in the summer when the sun is hot, but then it works of itself, and costs nothing. It is a well-known physical fact that glass is a trap for heat—that is, apparently it lets it in, but will not let it out again. To be more exact, it permits of the passage of luminous rays of heat, but not of opaque. The direct rays of the sun are luminous, but those which are radiated from a comparatively dull substance are opaque. For instance, in a greenhouse the heat is found to be very much more oppressive than in the hottest place outside. This principle is utilised in the solar extractor (Fig. 125). The appliance consists of a box formed with a sloping top like a desk, the top being glazed with a double thickness of glass as shown by Fig. 126. The dimensions may vary considerably, but those given in Fig. 125 will make a very useful and practicable size, namely, length, 20 in.; breadth, 12 in.; height at back, 12 in.; at front, 6 in.

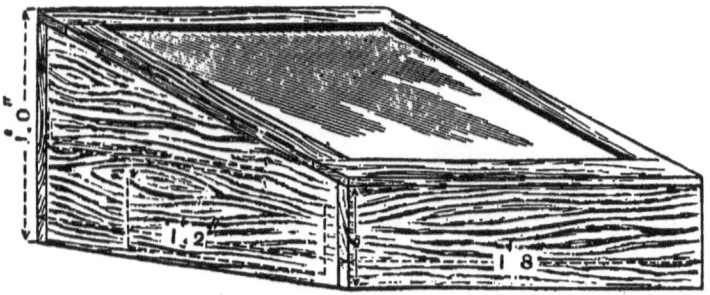

Fig. 125.—Solar Wax Extractor.

It should be made of very sound and dry stuff, preferably yellow pine, and it would be well to dovetail it together at the corners. The bottom should be grooved and tongued at the joint, or else made of one piece of wood. It would be a great improvement to line the whole structure with tinplate, which would ensure its being wax-tight. The top consists of a frame of 2 in. by 1 in. stuff mortised together at the corners, and rebated to take the glass, the rebate being $\frac{3}{4}$ in. by $\frac{1}{4}$ in. The glass is to be placed in a slight bedding of soft putty, and then a strip $\frac{3}{8}$ in. by $\frac{1}{4}$ in. is to be tacked to the frame close up to the glass; the other glass is then to be put in a similar bedding of putty, and another strip tacked on over all. The object is to have the glass air-tight in the frame, and this can easily be secured by a judicious use of putty or white lead. The frame may now be attached to the box by means of a couple of hinges at the back, and two hooks in front will keep it down close.

A tinplate shelf or tray is now to be made, the length of the inside, and approaching within an inch or so of the front. Three sides of this shelf

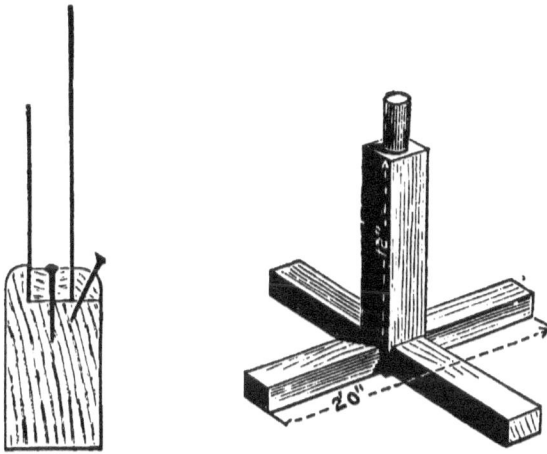

Fig. 126.—Glazing Top of Solar Wax Extractor.

Fig. 127.—Foot of Solar Extractor Stand.

are to be turned up for 1 in., as also the corners, to get a touch of solder. Tray supports are now to be affixed to the inside. If the box is lined with tinplate, these supports would take the form of pieces of tinplate soldered to the ends, and turned up at right angles, like L iron. If, however, the extractor is not lined, strips of wood tacked against the ends would do. The tray is to slope slightly forward so that the wax will run into the receptacle placed in front for it. The strips which support it will be placed about halfway up the ends.

Over the tray there is a sieve of tinned wire netting, bound with tin, on which the combs to be converted into wax are placed. This sieve is ¼ in. from the tray, supported with tinplate strips standing edgeways across it. The tinned edges should be turned up for ¼ in. to catch the comb and prevent it from slipping off.

The box to catch the wax underneath is as long as will fit between the tray supports, and may be as wide as the extractor, or any less width. It is made of tinplate with a wired top, and is, of course, water- and wax-tight. Care should be taken that the fluid wax will all flow into the receptacle placed for it, and not flow over its ends where it is not wanted.

A convenient stand for the solar extractor is

Fig. 128.—Revolving Top of Wax Extractor.

Fig. 129.—Washer and Screw.

shown by Fig. 127. To make it, get two pieces of wood 2 ft. long by 2 in. square, and halve them together in the middle. Now get another piece 15 in. long by 3 in. square, and round the upper end for a distance of about 3 in. to 1½ in. in diameter; then cut the lower part to fit over the junction of the cross pieces, and fix it to them with one long spike driven from underneath, and some smaller nails at the sides, having it at right angles to the cross pieces.

The revolving top to the stand (see Fig. 128) may next be taken in hand; it may be any convenient size, and about 3 in. thick. The most important item in its construction is the boring of the hole, which is 1½ in. in diameter, and should be exactly at right angles to the upper surface. It is countersunk on top, so that the washer and screw (Fig. 129) will be flush, or, if anything, somewhat lower than the surface of the wood. When the entire stand is put together, the extractor can be attached to it by means of four

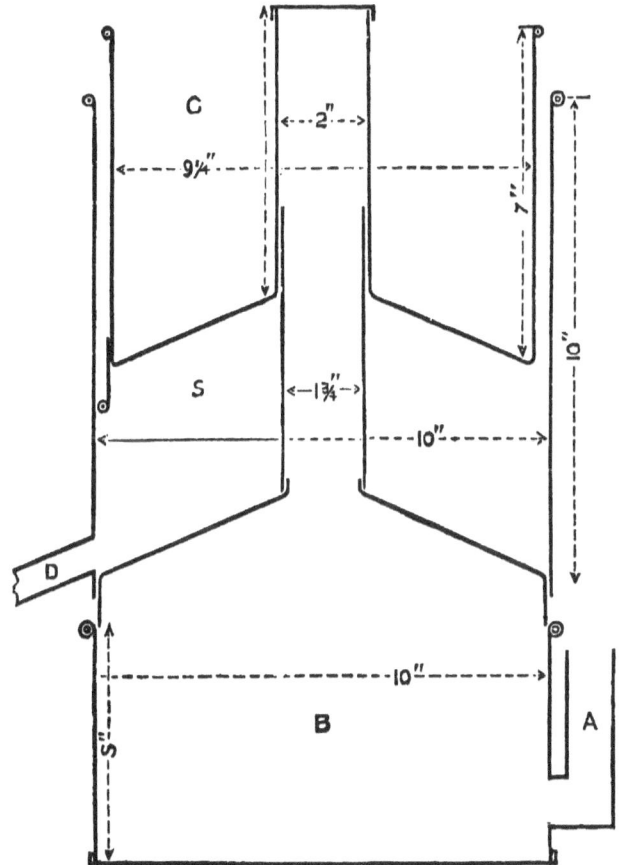

Fig. 130.—Section of Gerster Wax Extractor.

screws passing upwards into the bottom. The object of the revolving part is, of course, to enable the glass top to be turned, so as to catch the direct rays of the sun.

The Gerster wax extractor, of which a section

is given in Fig. 130, consists of four parts: (1) the boiler; (2) the steamer; (3) the comb basket; (4) the cover, or lid. It is hard to dish the covers to a nice curve without special tools and blocks; and, consequently, it is better to buy a saucepan cover for a few pence and make the other parts of any piece of apparatus to suit the cover. Supposing that the cover is 10 in. in diameter, the boiler and steamer can be made the same size.

The boiler can be taken in hand first. It is advisable to make it of copper, as it is then far more lasting. If, however, tinplate is used, on no account should acid be employed as a flux for the solder, as it would soon eat its way through the plates. The boiler may vary much in height, but 5 in., as shown in Fig. 130, is suitable.

The pipe A allows the height of the water in the boiler to be seen without taking off the steamer, which would be an awkward thing to do often. This pipe is about 1 in. in diameter, and has a cork or metal cap to cover the top. Instead of it a U-shaped piece of tin could be soldered over the hole in the boiler, and would be equally efficacious.

The steamer is of the same external diameter as the boiler. The bottom is in the form of a cone, of which a pattern is given by Fig. 131, and a rim of doubled tinplate is affixed to the lower edge, small enough to fit into the boiler. It is made just in the same manner as an ordinary vegetable steamer, which will be a good guide in its manufacture. The apex of the coned bottom has a hole cut in it, the edge is slightly turned up, and a $1\frac{3}{4}$-in. tinplate tube, 6 in. long, is soldered firmly to it. Another tube D is soldered into the side, as low down as possible, so that it will drain out all the contents of the steamer.

The comb basket is made of perforated tinplate, one hundred holes to the inch. The cylindrical part, $9\frac{1}{4}$ in. in diameter and 7 in. high, can first be

WAX EXTRACTORS.

made. It may be necessary to give a rule for cutting out the material for making hollow cylinders such as this: Multiply the diameter by 3¼, and add what is required to make the joint. If a plain overlapping joint is used, add the amount of lap; but if a hooked joint is used, three times the length of the turned over parts should be added, usually about ¾ in. In the present case, a strip of perforated tinplate, 29⅝ in. by 7⅝ in., will form

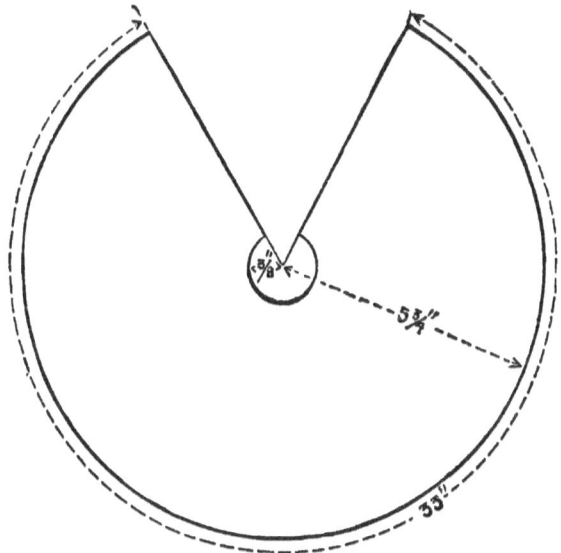

Fig. 131.—Pattern for Cylindrical Top of Boiler.

the cylinder, the extra width being ⅜ in. for the wiring on top, and ¼ in. to make the joint at the bottom.

The conical bottom of the basket has the same slope as that of the steamer; the same pattern will do for both, except that the radius of the basket pattern may be ⅜ in. less. A perforated tinplate tube, similar to that in the steamer, but ¼ in. larger, is fixed in the centre of the comb basket; but, while the tube in the steamer is open at both ends, that in the basket is closed on top with a piece of plain tinplate—the cover of a coffee canister does capitally.

Three legs of doubled tinplate are soldered, equidistant from each other, to the lower edges of the comb basket so as to keep the bottom 1 in. from that of the steamer. In Fig. 130 the three parts as drawn are separated from each other, but they would fit down into place in actual use.

To use the Gerster extractor, water is placed in the boiler, which is then put over a fire; the combs are smashed up and put in the comb basket, which is put into place, and the cover fits over all, and keeps in the steam. Presently, when the water begins to boil, the steam passes up through the centre tube of the steamer, hits against the closed top of the basket tube, and is disseminated through the combs, which it soon reduces to a fluid state. The wax and condensed steam run through the tube D into a vessel placed for their reception, while the dirt and refuse remain in the comb basket. In Fig. 130, A and D indicate the pipes; B the boiler; C the comb basket; and S the steamer.

The basket can be cleaned by a liberal application of hot water in which washing soda has been dissolved, and the point of a brush will take out any stubborn pieces of dirt.

All pieces of apparatus should be kept scrupulously clean.

CHAPTER XIV.

BEE KEEPERS' MISCELLANEOUS APPLIANCES.

This, the concluding chapter, will concern itself with a number of appliances more or less indispensable to the bee keeper.

A bee feeder which has two wooden floats is shown in plan by Fig. 132, and in vertical section

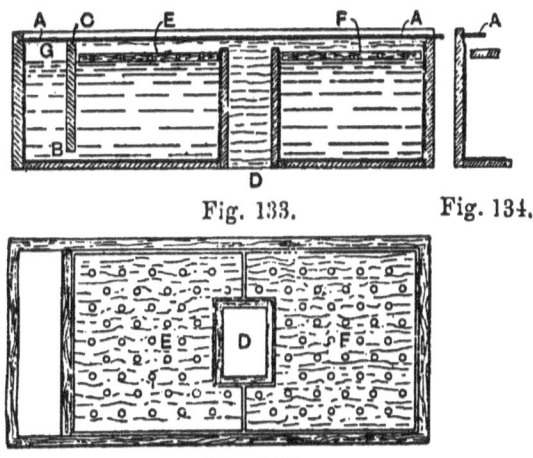

Fig. 132.—Plan of Float Bee Feeder. Figs. 133 and 134.—Sections of Bee Feeder.

by Fig. 133. First make a wooden box of any size up to 14 in. long by 8 in. wide by 4 in. deep. At a distance of $\frac{3}{16}$ in. from the top, run round the sides and one end a groove as shown at A (Figs. 133 and 134), into which the glass cover will slide freely; then at one end fix a partition C with ¼-in. clearance at the bottom B (Fig. 133) and up to the glass at the top. At D (Figs. 132 and 133) cut a hole about 3 in. by 2 in. through the bottom, and in this hole fix a wooden funnel, which must be kept ¼ in. below the glass at the top. The box must be carefully put together, so that it will be perfectly watertight, or a tin lining should be provided as shown by the inner line on Figs. 133 and 134, and if this is used the centre funnel should also have a further lining of perforated tin to enable the bees to gain a foothold. Zinc should not be used. The two floats E and F are made of ¼-in. stuff pierced with a number of $\frac{3}{16}$-in. holes. The bees pass up through D and over the top of the funnel on to the floats, and the feeder is filled

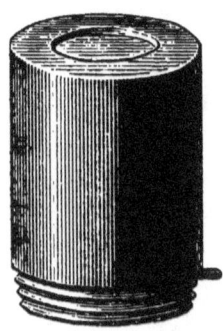

Fig. 135.—Raynor Bee Feeder.

Fig. 136.—Base for Bee Feeder.

by sliding back the glass cover and pouring in syrup at G.

A feeder of altogether different construction is shown by Figs. 135 to 137. This is the Raynor, one of the best for all round purposes. It consists of a bottle having a screw-on cap, which is perforated with a dozen holes in a semicircle in such a way that one or more can be brought over the circular slot which is shown in the stand (Fig. 136). A pointer soldered to the cap indicates the number of holes which are uncovered to the bees underneath.

Make the base of the feeders of hard wood,

BEE KEEPERS' MISCELLANEOUS APPLIANCES. 143

turned to the section shown in Fig. 137. It may be 6 or 7 in. across, and 1½ in. high. The recess in the top is made to fit the 2 lb. screw-top bottles, which can be bought from all dealers in bee appliances. The top of the dome is turned to about ⅛ in. thick, a circle marked round while in the lathe with the corner of the chisel, and the slot ⅛ in. wide, cut out with a narrow chisel and penknife.

The slot is to be no more than half a circle. The feeding bottle is then laid in place and holes pricked through its cap through the slot of the stand, with a darning-needle or fine awl. These

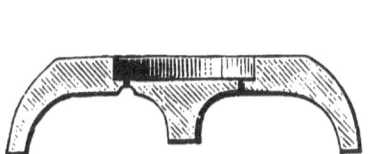 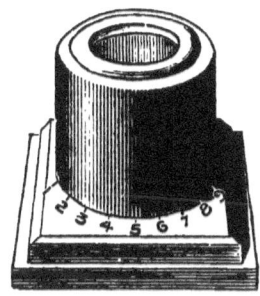

Fig. 137.—Section of Bee Feeder Base.

Fig. 138.—Bee Feeder with Square Base.

holes may be about a dozen in number, as may be inferred from Fig. 136. A tinplate pointer must be soldered to the cap, and numbers stamped on the stand corresponding to the number of holes open.

When the bottle is inverted, the syrup will not run out of the holes, owing to the air pressure and capillary attraction, but the bees can easily suck the syrup through them. The dome can be lined with cloth or chamois leather to keep it snug, but this is not an essential.

When a stand is required at a moment's notice, use a piece of pine 5 in. square, and tack a slip 1 in. wide and ⅜ in. thick all round, as in Fig. 138.

Then cut a hole right through the top large enough to take a tin canister cover of the correct size, and then flange this over and secure it in place with a couple of tacks. Cut the slot in the tin and make the holes in the bottle cap as before, and the work is done. A pointer, of course, will be needed as before.

The Hone dummy feeder is shown in section by

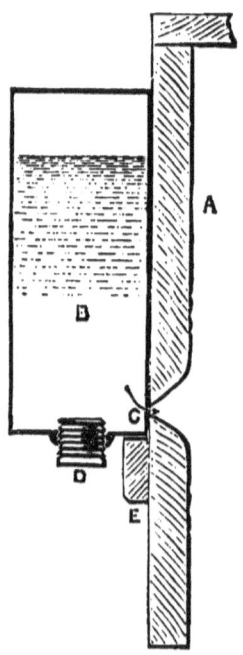

Fig. 139.—Section of Hone Dummy Feeder.

Fig. 139. A slit is cut in the dummy A to enable the bees to take the syrup, which is contained in an oblong tin box B with perforated edge placed at the back. The slot in the dummy is 6 in. long, and the tin box 8 in. by 5 in. by 2 in.; twelve holes C are made along the lower edge opposite the slit, and a screw plug D and leather washer keep the hole through which the tin is filled perfectly tight. The

BEE KEEPERS MISCELLANEOUS APPLIANCES. 145

tin is kept in place by two pieces of wood 2 in. by 5 in. by ¾ in., nailed edgeways to the dummy, and two other pieces 5 in. by 1 in. by ½ in., nailed to the backs of these, so that ¼ in. embraces the back of the tin. The whole works like the female portion of a slide. A small strip E underneath prevents the tin from going too low. If the screw cap is an objection, it could be replaced by a tube and good cork, an indiarubber stopper being still better. For spring feeding, when only a small quantity of syrup is required to be given at a time, some of the holes in this feeder may be plugged up with wax.

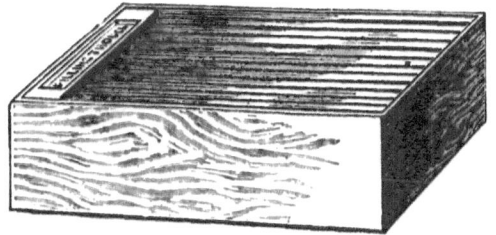

Fig. 140.—Rapid Bee Feeder.

The principal objection to the feeders shown by Figs. 135, 138, and 139 is that they must be frequently attended to, on account of the small quantity of syrup which they contain. This defect has been overcome in the American rapid feeders, of which there are many in the market. The float feeder, the first appliance described in this chapter, is also free from this defect.

Fig. 140 is a general view and Fig. 141 a cross section of one of these American rapid feeders. It consists of a trough holding about 10 lbs. of syrup. This is supported over the brood nest by the ends and a pair of supplementary sides, which allow the bees free access to the top of the trough without permitting them to escape, a thin board acting as a cover. The vertical lines in Fig. 141 in-

J

dicate a sort of ladder made of very thin wood, its object being to prevent the bees being drowned in the trough.

To make this feeder, two pieces of clean pine 10½ in. by 4¾ in. by 1 in. are prepared for the ends. Rebates 1⅛ in. wide by ⅜ in. deep are then cut round three sides of each. Two pieces 14 in. by 3¼ in., and one 14 in. by 9¼ in. by ¼ in., are got for the sides and bottom of the trough. This may be now completed by nailing the sides and bottom to the rebated parts of the ends forming a trough 8¼ in. wide and 3¼ in. deep, and with the ends projecting ⅜ in. beyond the sides and bottom. The outer sides, which are 14¾ in. by 4¾ in. by ½ in., are now nailed to the projecting parts of the ends, which will leave a space of ⅜ in. between the sides of the trough and these outer sides. If the whole structure be now laid upon a table, it will be found that the bottom of the trough is ⅜ in. from the surface of the table. A partition is now placed 1 in. from the end of the trough to form a filling space. This partition is pierced with ⅛ in. holes, so that when the syrup is poured into the smaller compartment, it will run into the larger, which can thus be filled without removing its cover. The cover is made of ¼ in. stuff of such a size as to cover the larger compartment, that is, 10½ in. by 13 in., and two little cleats are put on to prevent it from warping.

If an examination is now made, it will be found that the bees can crawl up between the inner and outer sides of the feeder and over the side of the trough into the food. A careful examination at this stage will show the course the bees will take; and if any passage is less than ⅜ in., it should be enlarged. The top edge of the trough will evidently be ⅜ in. lower than the sides of the feeders and the under surface of the cover. The bees could at this stage find their way into the smaller or filling compartment through small spaces which communicate with the hives. These are now stopped up with scraps of wood tacked over them, and a long narrow strip of glass is cut to cover the compartment.

The ladder (see Fig. 141) which enables the bees to take the food without the risk of being drowned can now be made. The best material to employ is the wooden dividers used to separate the sections in a section crate. They can be cut 12 in. by 3⅛ in., and seventeen or eighteen of them will be required. Twice as many pieces of wood ⅜

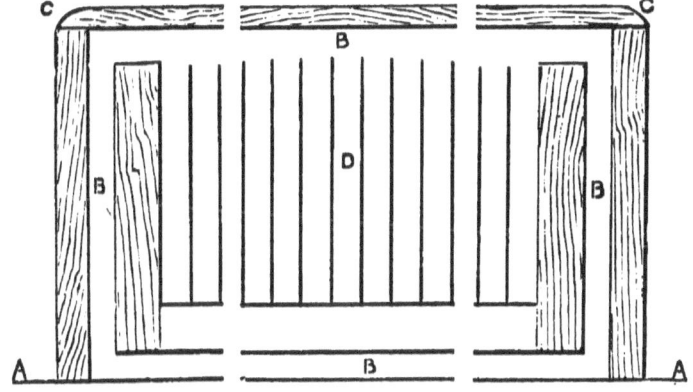

Fig. 141.—Section of Rapid Bee Feeder.

in. thick by about ¾ in. by 1¼ in. will also be required. One of the thin dividers is then taken and marked with a pencil 3½ in. from each end. A couple of the small pieces of the wood are then laid on these marks and equidistant from the edges of the divider; another divider is then laid on top, and a tack through each thick piece secures the three together; another couple of thick pieces are put next, again a thin one, and the tacking continued as before until the pile is high enough to fit the breadth of the trough. The thin wood dividers of which this ladder is made are kept ⅜ in. apart by the little blocks between. The out-

148 *BEEHIVES AND BEE KEEPERS' APPLIANCES.*

side dividers will, when in place, be ⅜ in. from the sides of the trough, being kept so far away by similar blocks. To prevent the ladder from floating in the syrup, a couple of little wooden buttons attached to the sides of the trough can be turned over it, while two strips ⅛ in. thick are tacked to the bottom of the trough on the inside to enable the syrup to flow freely to every part. This is an excellent feeder. It takes the place of the section crate, but of course is only suitable for use in the autumn, when stocks have to be fed up rapidly before the winter.

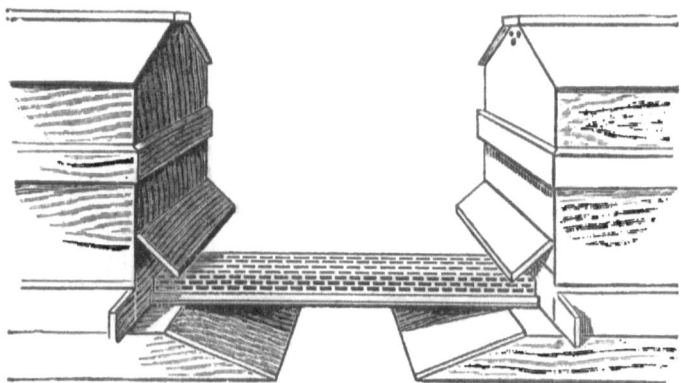

Fig. 142.—Bennett's Self-hiver.

A self-hiver for conducting a surplus swarm from its old quarters to a new hive is of the greatest advantage. That shown by Fig. 142 is known as the Bennett, and its position with reference to the hives must first be understood. One of the hives shown contains the stock of bees which is expected to swarm, while the other is the empty hive containing frames, foundation, quilts, and possibly comb and honey, into which it is desired to lead the swarm. The inventor has described as follows how this useful appliance is made:— First get a thin board of ½ in. stuff 2 ft. long and 6 in. wide (any size can be adopted at the dis-

BEE KEEPERS' MISCELLANEOUS APPLIANCES. 149

cretion of the maker). Obtain also a piece of queen excluder zinc, same length as the board, and 9 in. wide, and bend 1¾ in. on each side, along the whole length of the zinc, and tack the bottoms of the bent sides along the edges of the board. Nail along each of these sides a thin strip of wood ¾ in. wide, so forming a sort of square tunnel— 2 ft. long, 6 in. wide, and 1¼ in. deep, with both ends open, and a sort of miniature alighting board along its sides. Place an empty hive in front of the one expected to swarm, draw apart the slides to form an entrance 6 in. wide, and put the cage

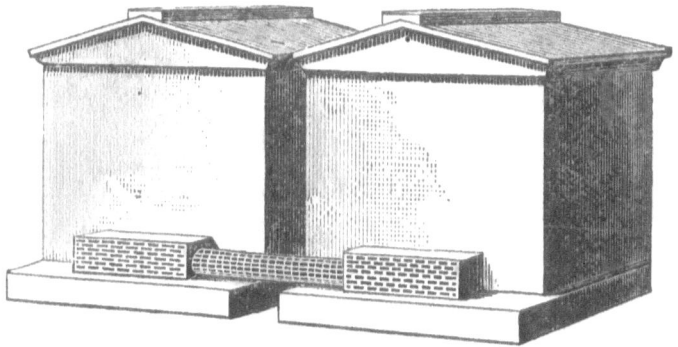

Fig. 143.—Alley's Self-hiver.

or tunnel on the entrance board of each hive, the open ends of the cage being in front of the entrances of both hives.

With regard to the practical working of the Bennett self-hiver, one bee keeper reports that it did splendidly, the swarm settling down quietly in its new home. In another case the swarm went off, leaving the queen in the tunnel vainly trying to follow. When she found that she could not leave, she returned to the parent hive, and the swarm joined her there. When the same swarm issued again, and the queen tried to get through the excluder zinc, the owner removed the empty hive and tunnel, and stopped up the open end of

150 *BEEHIVES AND BEE KEEPERS' APPLIANCES.*

the latter with paper, so that the queen had perforce to enter the empty hive. The swarm soon joined her there, and took to their new quarters readily.

Alley's self-hiver (Fig. 143) is on the same principle, the only difference being that the tunnel leads to another hive at the side of the swarming one, instead of in front. It does not promise,

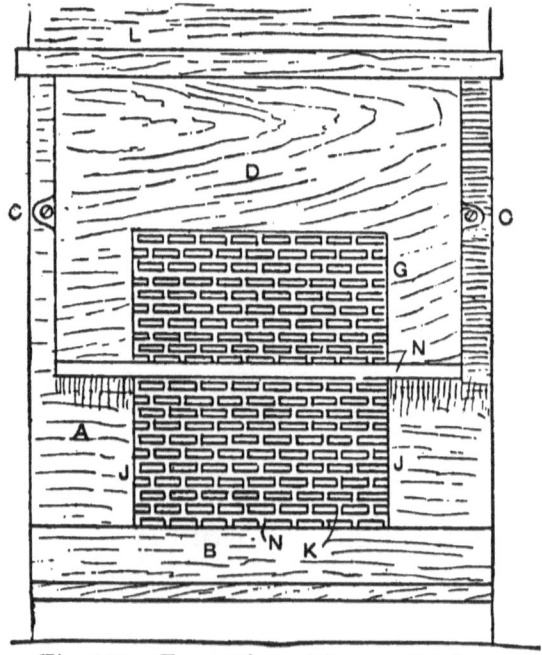

Fig. 144.—Front View of Swarm Catcher.

however, as well as Bennett's, there being so many corners and angles in it.

A swarm catcher for a beehive is shown in front view by Fig. 144, and in section by Fig. 145, in which A is the beehive, the swarm catcher being attached to it by means of two iron plates with screws at C. B is the alighting board of the hive. The swarm catcher D consists of a box made to

BEE KEEPERS' MISCELLANEOUS APPLIANCES. 151

take three standard frames E (Fig. 145), which are fitted with full sheets of brood foundation. The

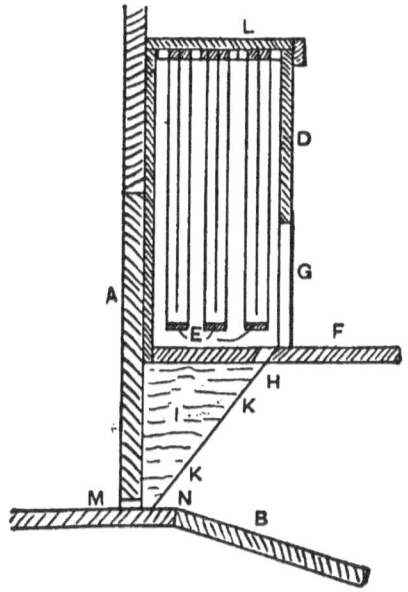

Fig. 145.—Section of Swarm Catcher.

bottom of the catcher is extended about 3 in. in advance of the front, as shown at F (Fig. 145), to form an alighting board for the swarm. A hole

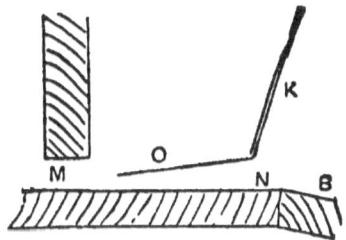

Fig. 146.—Hive Entrance with Flexible Springs.

G about 6 in. by 4 in. is cut in the front of the catcher, and is covered with queen excluder zinc, and a slot, 7 in. long by ½ in. wide, is cut in the bottom at H. Two triangular pieces I (Fig. 145) are cut to the shape shown, and nailed to the bottom of the catcher on each side at J (Fig. 144), and the space between them is covered with excluder zinc K (Figs. 144 and 145). A lid L is fitted to the top of the catcher to keep in the swarm and to keep the bees dry in case of rain.

The method of working this catcher is as follows:—When the swarm issues through the flight hole M, the workers pass through the excluder zinc K, but as the queen cannot pass this,

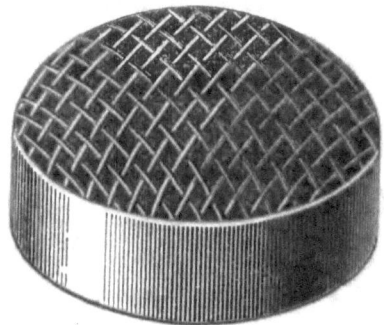

Fig. 147.—Pipe-cover Queen Bee Cage.

she walks up K and passes through H into the catcher, where the bees forming the swarm join her. In the evening the parent hive is moved about a yard away, and the combs in the catcher are put into a new hive with the swarm, together with four more combs, one of which should, if possible, contain brood.

In some swarm catchers the bottom slots of the excluder zinc N (Figs. 144 to 146) are cut away, and a number of very flexible brass springs are fixed across the entrance, to enable the workers to enter the hive without hindrance when returning home loaded with pollen. The arrangement of these will be clear from Fig. 146, in which M

BEE KEEPERS' MISCELLANEOUS APPLIANCES. 153

indicates the entrance to the hive, K the excluder zinc, B the alighting board, O the springs, and N the entrance under the excluder zinc.

Queen cages are often a necessity; the simplest of them is, perhaps, the pipe-cover queen cage, shown by Fig. 147. It can be made as follows:—

Get a strip of tinplate 6½ in. long and 1 in. wide; make it into a ring and solder the ends together. Obtain a circular piece of wire gauze or

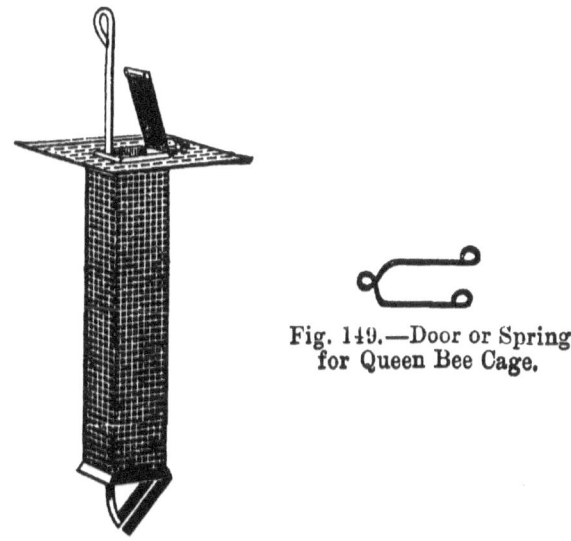

Fig. 149.—Door or Spring for Queen Bee Cage.

Fig. 148.—Another Type of Queen Cage.

perforated tinplate, 2 in in diameter, and solder it on as a top, and the cage is complete.

A disadvantage of this cage is that the bees must be disturbed in releasing the queen. This difficulty is overcome in the cage shown by Fig. 148. It consists of a rectangular cage, formed of perforated tinplate or wire gauze. Its dimensions are: Length, 4 in. or 5 in.; width, 1¼ in.; thickness, ½ in. A piece of metal, 3¾ in. wide and as long as the cage, bent over a piece of wood 1¼ in. by ½ in., will just make it. A flange is then made

154 *BEEHIVES AND BEE KEEPERS' APPLIANCES.*

of plain tinplate for the top. This may be about 2½ in. by 1½ in., having a hole 1¼ in. by ½ in. in its centre, into which the cage fits and is soldered, leaving a little bit projecting at the back, which is turned over a pin and acts as a hinge for the top door. This may now be made of tinplate 1½ in. by ¾ in.—a hinge formed at the back by turning up the tinplate over a fine piece of wire or a pin, and fitted to the part of the cage projecting above the flange.

A door for the bottom can be made of wire, bent into the form of Fig. 149, the two wires being ⅛ in. apart, and the distance being 1¼ in. from centre to centre of the loops, that is, from the single loop at the left to the line joining

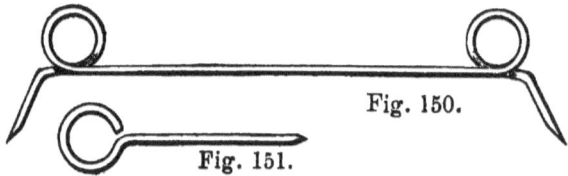

Figs. 150 and 151.—Driving Irons.

the pair of loops to the right. This door may now be fastened to the cage by passing a pin through the lower right-hand corner of the cage, so as also to pass through the two loops of the door. Some provision should be made, either by washers or a couple of twists of fine wire, to keep the door in the centre of the cage; otherwise it might move against the side, and allow the bees to have access to the imprisoned queen. A light wire hooked on to the front loop, and passing with slight friction through a hole in the flange, completes the cage. The queen is released by pressing down this wire, which projects about an inch above the flange, and ends in a loop.

Driving irons are shown by Figs 150 and 151. It would be advisable to make a great number of

BEE KEEPERS' MISCELLANEOUS APPLIANCES. 155

sets, as they are easily lost. They are made of wire nearly ¼ in. thick; the loops in Fig. 150 are 9 in. apart, and the ends about 1½ in. long, roughly pointed with a file. About 20 in. of wire are required for the form shown by Fig. 150, and two

Fig 152.—Bingham Knife.

like that and one like Fig. 151 complete a set; the latter is about 9 in. long, and the loop 2 in. diameter. It requires 15 in. of wire.

The making of a Bingham knife (Fig. 152) is not to be undertaken, except the manufacturer has special facilities for that kind of work. The blade is made of good saw steel, 6 or 7 in. long and from 2 in. to 3 in. wide, shaped to the pattern shown, and sharpened all round like a chisel, from the under side only. A tang is made of ⅜ in. iron or steel, one end being pointed to fit the handle and the other flattened so as to be riveted to the blade. The two rivets should be countersunk into the under side of the blade and ground flush with its surface; when the blade is laid flat on a board,

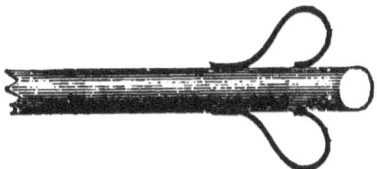

Fig. 153.—Comb Cutter.

the handle is raised about an inch from it, and the tang should be bent in such a manner as to secure this. The knife is not unlike a mason's trowel.

An appliance for cutting passages through the

combs on the approach of winter is illustrated by Fig. 153. It is simply a tinplate cylinder about 1 in. in diameter and 4 in. or 5 in. long, having one end serrated so as to cut the comb more easily. Near the other end two wire lugs are soldered to afford a better hold to the fingers, as shown in Fig. 153.

The Cheshire transferring board, shown by Fig. 154, affords facilities for transferring the combs cut from a skep to a bar-framed hive. The table proper consists of sixteen tongues projecting from a back support about 10 in. To make it, a piece of wood about 16 in. long, 3 in. wide, and 1 in.

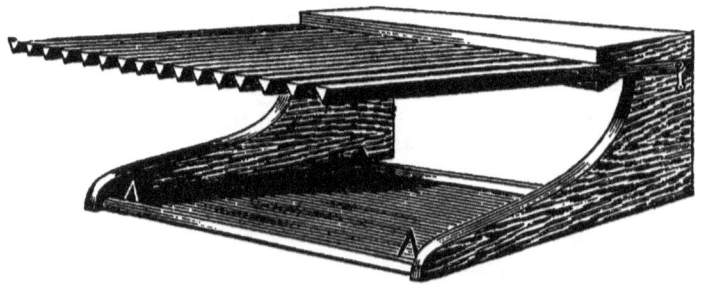

Fig. 154.—Cheshire Transferring Board.

thick is planed quite flat and out of winding, the under surface being especially true. The tongues are all cut out of a piece of yellow pine 13 in. long, 1 in. thick, and about 11 in. wide. Fig. 155 is a view of the end of this piece of wood, showing how the tongues may be cut out. Of course, it will require care to cut the bevels to the proper angle, but any want of accuracy in the saw can be rectified by the trying plane. The dimensions of each tongue, when finished, will be: length, 13 in.; top width, $\frac{3}{4}$ in.; bottom width, $\frac{1}{4}$ in.; depth, 1 in. The tongues must be nailed or, preferably, screwed to the back piece, each tongue being at right angles to the back, and the edges $\frac{1}{4}$ in. apart. It is desirable to fill up the spaces between the tongues immediately underneath the support of the back with pieces of wood nicely fitted in.

The legs may be either fixed or folding, but in any case they will be cut to the shape shown in Fig. 154, the height being 6 in. or 8 in.; length from front to back, 12 in.; and thickness of the wood, 1 in. If they are folding, hinges should secure them to the back support, strap or butt hinges 3 in. wide doing the business very well. The outside surface of the legs is quite flush with the outer edge of the last tongue, and to prevent

Fig. 155.—Cutting Tongues from Board.

the leg shutting up when not wanted, a hook and eye, such as is used to hold the first door of a cupboard, is fitted, the hook being secured to the leg and the eye screwed into the surface of the tongue. This is shown at the right-hand side towards the back of Fig. 154. A tinplate tray to fill the space between the legs is necessary to catch any honey which may drop from the combs during manipulation. As the transferrer will be always exposed to the smearing of honey, it would be well to give it several coats of a hard varnish, which will make the surface washable. The bee keeper should always aim at perfect cleanliness.

INDEX.

Abbot's Broad-shouldered Frame, 49
Alley's Self-hiver, 150
American Jointed Frame, 52
—— Rapid Feeder, 145

Bar-frame Beehive, 9—18
—— —— Brood Chamber, 14
—— ——, Floor Board, 13
—— ——, Furnishing, 48
—— ——, Lift, 16
—— ——, Materials for, 26
—— ——, Porch, 15
—— ——, Roof, 16
—— ——, Tiering, 25—40 (*for details see* Tiering Bar-frame Beehive)
—— —— Ventilation, 17
Basket, Comb, for Gerster Extractor, 138
B.B.K.A. Standard Frame, 49
Bee Escape, Cone, 17
—— ——, Porter, 98
—— Feeder, American, 145
—— ——, Float, 141
—— ——, Hone Dummy, 144
—— ——, Rapid, 145
—— ——, Raynor, 142
—— Smokers, 102—115
"Bee-space," 11
Bellows for Bee Smoker, 108
—— Springs, Bee Smoker, 108
Bennett's Self-hiver, 148
Bingham Bee Smoker, 102
—— Knife, 155
Blast Pipe for Clarke Smoker, 115
—— ——, Cone, 107
Block for Wiring Frame, 54
Body-box of Queen Bee Hive, 91—92
—— —— Tiering Bar-frame Beehive, 31
—— —— W.B.C. Beehive, 44
Boiler of Gerster Wax Extractor, 138
Bolt for Cross-bar, 131
Bracket, Porch, 39
Broad-shouldered Frames, 50
Brood Chamber, 14
—— —— Lining, 17
—— —— for Observatory Beehive, 69

Brood Chamber for W.B.C. Beehive, 44
"Brood Nest," 12

Cages, Queen, 152—154
Case, Beehive Inspection, 84—88
Catcher, Swarm, 150—152
Chamber, Brood, 14
——, ——, Lining of, 17
——, Surplus, 12
Cheshire Transferring Board, 156
Clarke Smoker, 112
—— —— Blast Pipe, 115
—— —— Diaphragm, 114
—— —— Fire-box, 113
—— —— Funnel, 114
Cleaning Basket of Gerster Extractor, 140
Clearers, Super, 96—101
Comb Basket for Gerster Wax Extractor, 138
—— Baskets, 128
—— Cutter, 153, 154
Combination Hives, 11
Coned Blast Pipe, 107
Cones, Bee, 17, 96, 97
Crates, 65
Crossbar Bolt, 131
Cylinder Honey Extractor, 123

Diaphragm of Clarke Smoker, 114
Divider, Section, 66
Driving Irons, 154
"Dummy," 61, 62
—— Feeder, 144
—— for Queen Bee Hive, 94

"Eke" for Increasing Height of Lift, 30
—— —— W.B.C. Beehive, 45
Embedder, 56
—— made with Floor Brad, 57
——, Wolblet Spur, 57
End Spacers, Staples used as, 61
—— ——, Cast Metal, 59
—— ——, Howard, 59
—— ——, W.B. Tinplate, 59
Entrance Slides, 39
Escape, Cone, 17, 96

INDEX.

Escape, Porter, 98
Excluder, Queen, 63
Extractor, Cylinder, 123
——, Gerster, 133
——, Honey, 116—132
——, Little Wonder, 116
——, Solar, 133
——, Wax, 133—140

Feeder, American Rapid, 145
——, Float, 141
——, Hone Dummy, 144
——, Rapid, 145
——, Raynor, 142
Feeding Syrup, 67
Fire Box of Clarker Smoker, 113
Float Bee Feeder, 141
Floor Bearer for Tiering Bar-frame Beehive, 35
—— Board, 13
—— ——, Observatory Beehive, 68
—— ——, Tiering Bar-frame Beehive, 34, 45
—— Brad, Embedder made with, 57
—— for Tiering Bar-frame Beehive, 35
Folding Section, 65
Foundation, Fixing, in Frame, 54
——, Section with, 64
Frame-box for W.B.C. Beehive, 44
Frames, Abbot's Broad-shouldered, 50
——, American Jointed, 52
——, B.B.K.A., 49
——, Block, for Wiring, 54
——, Broad-shouldered, 50
——, Fixing Foundation in, 54
——, Handle for Lifting, 88
——, Hoffman Self-spacing, 59
——, Hooks for Lifting, 88
—— for Observatory Beehive, 72
——, Self-spacing, 59
——, Shoulderless, 50
——, Standard, 49
——, Wax Sheet fixed in, 54
—— with W.B.C. Metal Ends, 50
Funnel for Bingham Smoker, 103
—— —— Clarke Smoker, 114
Furnishing Beehive, 48—66

Gauge-board, 56
Gerster Extractor, 133
—— —— Boiler, 138
—— —— Comb Basket, 138
—— —— Steamer, 138

Gerster Extractor, Using, 140
Guide-board, 56

Handle for Lifting Frames, 88
Hatchet Stake, 103
Hoffman Self-spacing Frame, 59
Hone Dummy Feeder, 144
Honey Extractor, Cylinder, 123
—— ——, Little Wonder, 116
——, Marketing, 64
Hooks for Lifting Frames, 88
Howard Tinplate End, 59

Inspection Case for Beehives, 84—88
Irons, Driving, 154

Joint, Tongue-and-groove, 34

Knife, Bingham, 155

Legs for Stand of W.B.C. Hive, 45
Lift, Beehive, 16
—— for Observatory Beehive, 72
—— —— Tiering Bar-frame Beehive, 36
—— —— W.B.C. Beehive, 43
Little Wonder Honey Extractor, 116
Lining of Brood Chamber, 17

Makeshift Hives, 19—24
Marketing Honey, 64
Metal Ends for Frames, 50, 59
Mounting Observatory Beehives, 83

Nest, Brood, 12
"Neucleus" Hives, 20

Observatory Beehive Brood Chamber, 69
—— —— Floor-board, 68
—— —— Frames, 72
—— —— Internal Frame, 74
—— —— Interior Fittings, 72
—— —— Lift, 72
—— ——, Mounting, 83
—— ——, Permanent, 69—73
—— —— Porch, 72
—— —— Quilt Covering, 73
—— —— Roof, 72
—— ——, Stocking, 80
—— —— Super, 73
—— ——, Temporary, 74—83

Pine's Cast Metal End, 49
Pipe, Cone Blast, 107
Pipe-cover Queen Bee Cage, 152
Porch for Beehive, 15
—— Brackets, 39
—— for Observatory Beehive, 72
—— —— Temporary Hives, 23

Porch for W.B.C. Beehive, 41
Porter Bee Escape, 98

Queen Bee Cages, 152—154
—— —— Rearing Hive, 89—95
—— Excluder, 63
Quilts, 64, 73

Rapid Bee Feeder, 145
Raynor Bee Feeder, 142
Regulating Slides, 17
Roof for Beehive, 16
—— Gables, Beehive, 37
—— for Observatory Beehive, 72
—— —— Temporary Hive, 22
—— —— Tiering Bar-frame Beehive, 37
—— —— W.B.C. Beehive, 43
—— Wings, Beehive, 38

Section Divider, 66
——, Folding, 65
—— with Foundation, 64
Sections, Crate for, 65
Self-hiver, 148
——, Alley's, 150
——, Bennett's, 148
Self-spacing Frame, 59
Slides, Entrance, for Tiering Bar-frame Beehive, 39
——, Regulating, 17
Smoker, 102—115
——, Bellows for, 108
——, Bingham, 102
——, Clarke, 112
—— Diaphragm, 106
Solar Wax Extractor, 133
Spacing Frames, 60
Spring for Queen Bee Cage, 154
Springs for Bellows of Bee Smokers, 108
Spur Embedder, Woiblet, 57
Stake, Hatchet, 103
Stand for Queen Bee Hive, 94
—— —— W.B.C. Hive, 45
Standard Frame, 49
Staples Used as End Spacers, 61
Steamer for Gerster Wax Extractor, 138
Stocking Beehive, 66, 67
—— Observatory Beehive, 80
Super-clearers, 96—101
"Supers," 12
—— for Observatory Beehives, 73

"Supers," Removing, 101
—— for W.B.C. Beehives, 44
Surplus Chambers, 12
Swarm Catcher, 150—152
Syrup, Feeding, 67

Temporary Beehive, 19—24
—— —— Porch, 23
—— —— Roof, 22
—— Observatory Beehive, 74—83
Tiering Bar-frame Beehive, 25—40
—— —— —— Body-box, 34, 35
—— —— —— Entrance Slides, 39
—— —— —— Floor-bearer, 35
—— —— —— Gables, 37
—— —— —— Lift, 36
—— —— —— Main Floor, 35
—— —— ——, Materials for, 26
—— —— —— Roof, 37
—— Hives, 11
Tinplate End, W.B.C., 59
Tongue-and-groove Joint, 34
Transferring Board, 156

Unicomb Beehive, Temporary, 74

Ventilation of Beehive, 17

Wax Extractors, 133—140
—— ——, Gerster, 133, 137
—— ——, Solar, 133
—— Sheet, Fixing, in Frame, 54
"W.B.C." Beehive, 41—47
—— —— Body-box, 44
—— —— Brood Chamber, 44
—— —— Eke, 45
—— —— Frame-box, 44
—— —— Lift, 43
—— —— Parts, 41
—— —— Porch, 41
—— —— Roof, 43
—— —— Stand, 43
—— —— Super, 44
—— Metal Ends, 50, 59
Wings, Beehive Roof, 38
Wiring Frame to receive Foundation, 55
—— —— Block, 54
Woiblet Spur Embedder, 57
Wood for Beehive Construction,

www.ingramcontent.com/pod-product-compliance
Lightning Source LLC
Chambersburg PA
CBHW030325080526
44584CB00012B/717